Praise for Andy Clark's

THE EXPERIENCE MACHINE

"Clark is a leading figure in understanding the brain as a prediction machine. . . . This thoroughly readable book will convince you that the brain and the world are partners in constructing our understanding."

—Sean Carroll, *New York Times* bestselling author of *The Biggest Ideas in the Universe: Space, Time, and Motion*

"Is the universe a simulation? Yes! But the simulation takes place in your brain. In this engaging and fascinating book, Andy Clark explains how our expectations dominate the input of our senses to construct our individual perceptions of reality. After reading it, you'll look at human experience in a new way."

—Leonard Mlodinow, bestselling author of *Emotional* and *Subliminal*

"Marvelous. . . . [Clark reveals] your brain's mind-bending (and mind-making) predictive powers that construct the reality you see, hear, and feel. Without them, there is only buzzing, blooming confusion. Strap on your seatbelt and prepare to be amazed!"

—Lisa Feldman Barrett, author of *How Emotions Are Made* and *Seven and a Half Lessons About the Brain*

"Rare among science books, this one has changed the way I experience the world. I now feel the experience machine doing its work as I pay attention, am surprised, or catch myself having made completely ridiculous predictions. It's a book that will help you understand the way you see, think, and act—and it is also a pleasure to read."

—Susan Blackmore, author of *Consciousness* and *The Meme Machine*

"Fascinating and profound. . . . This book sets the highest possible bar for popular science writing and is sure to become an instant classic of the genre."

—David Robson, author of *The Expectation Effect*

"If you would like to read the most promising theory of how your brain works, told by the clearest and most colorful storyteller, this is the only book you need."

—Moshe Bar, author of *Mindwandering: How Your Constant Mental Drift Can Improve Your Mood and Boost Your Creativity*

"Remarkable. . . . A startling, profoundly illuminating account of our mind's predictive abilities."

—*Kirkus Reviews* (starred review)

"[An] eye-opening study. . . . The mind-bending research upends conventional wisdom about how humans interact with the world around them, and the lucid prose ensures lay readers won't get lost. This head trip delivers."

—*Publishers Weekly*

Andy Clark

THE EXPERIENCE MACHINE

Andy Clark is a professor of cognitive philosophy at the University of Sussex in the United Kingdom and at Macquarie University in New South Wales, Australia. He is the author of several books, including *Surfing Uncertainty: Prediction, Action, and the Embodied Mind* and *Being There: Putting Brain, Body, and World Together Again*. He is a fellow of the Royal Society of Edinburgh and the British Academy and an occasional academic consultant for Google U.K. He was principal investigator on a European Research Council Advanced Grant ("Expecting Ourselves: Embodied Prediction and the Construction of Conscious Experience") exploring the relationship between embodiment, prediction, and consciousness. His current research interests include embodied and extended cognition, robotics, computational neuroscience, and computational psychiatry.

The Experience Machine

HOW OUR MINDS PREDICT AND SHAPE REALITY

Andy Clark

VINTAGE BOOKS
A DIVISION OF PENGUIN RANDOM HOUSE LLC
NEW YORK

FIRST VINTAGE BOOKS EDITION 2024

The Library of Congress has cataloged the Pantheon Books edition as follows:
Names: Clark, Andy, [date] author.
Title: The experience machine: how our minds predict and shape reality / Andy Clark.
Description: First edition. | New York: Pantheon Books, [2023].
Identifiers: LCCN 2022037742 (print) | LCCN 2022037743 (ebook)
Subjects: LCSH: Cognitive science. | Experience. | Expectation (Psychology) | Neurosciences.
Classification: LCC BF311 .C5399 2023 (print) | LCC BF311 (ebook) | DDC 153—dc23
LC record available at https://lccn.loc.gov/2022037742
LC ebook record available at https://lccn.loc.gov/2022037743

Vintage Books Trade Paperback ISBN: 978-0-525-56725-7
eBook ISBN: 978-1-524-74846-3

vintagebooks.com

Printed in the United States of America
10 9 8 7 6 5 4 3 2 1

FOR ALEXA AND GILL, PRECISELY

What you see and what you hear depends a great deal
on where you are standing.
It also depends on what sort of person you are.

—C. S. LEWIS, *The Magician's Nephew*

Contents

PREFACE:
SHAPING EXPERIENCE

A FEW years ago, I was invited to speak at a popular science festival held in London. I'm a professor of cognitive philosophy (an odd title that reflects a rather eclectic set of interests spanning philosophy, neuroscience, psychology, and artificial intelligence) and I was about to give a talk on one of my favorite topics—the human brain as a "prediction machine." The festival, run by a popular science magazine, was called New Scientist Live. Every year, *New Scientist* (the magazine) invites experts in many different fields to give public presentations. This year, it was held in the huge ExCel center in London's docklands. Entering the ExCel center was like arriving at multiple conjoined ocean liners each hosting a different large-scale event. As a university professor, I'm no stranger to public speaking. But standing backstage at one of the larger auditoriums and thinking about the packed audience behind the curtain, I couldn't help but get the jitters. Maybe I should have made some last-minute changes to my slides. Maybe I ought to have worn a less startling shirt. Was there someone I forgot to thank? Suddenly, my anxious train of thought was interrupted by my phone buzzing in my pocket.

But my phone was not in my pocket. As I quickly remembered, not only had I removed it and placed it under the podium, I had also set it safely on airplane mode for the entire event. But buzzing I had felt—and clear, strong buzzing too. What I had experienced was a thoroughly modern phenomenon, a remarkably common trick of the mind now known as "phantom vibration syndrome." Given that I am a chronic long-term phone user, my brain has slowly come to expect the frequent intrusion of pocket-buzz, and I'm not the only one. A 2012 study found that 89 percent of college undergraduates reported feeling phantom phone vibrations, and it's been found to be particularly prevalent among medical interns, where fake buzzing is strongly associated with stress.* In 2013, the term was rated "word of the year" by Australia's *Macquarie Dictionary*.

It was fitting that these phantom vibrations should intrude just as I was about to launch my presentation. For although such phenomena are well known within psychology and neuroscience, they now fall into place as part of a much grander theory, one that I have been helping construct for the past decade. According to that overarching theory (the topic of my talk) phantom vibrations are just one vivid demonstration of the way all human experience is built. According to the new theory (called "predictive processing"), reality as we experience it is built from our own predictions. It was my habitual expectation of pocket-buzz that, combined with the stress of the occasion, created a clear buzzing sensation out of whole cloth.

Predictive processing speaks to one of the most challenging questions in science and philosophy—the nature of the relationship between our minds and reality. The theory, which has been steadily gaining momentum, changes our understanding of this relationship in ways that have far-reaching implica-

* All references, evidence, and supporting materials are gathered in the endnotes at the back of the book, where they are arranged by the relevant page number and a short identifying extract from the text.

tions. Contrary to the standard belief that our senses are a kind of passive window onto the world, what is emerging is a picture of an ever-active brain that is always striving to predict what the world might currently have to offer. Those predictions then structure and shape the whole of human experience, from the way we interpret a person's facial expression, to our feelings of pain, to our plans for an outing to the cinema.

Nothing we do or experience—if the theory is on track—is untouched by our own expectations. Instead, there is a constant give-and-take in which what we experience reflects not just what the world is currently telling us, but what we—consciously or nonconsciously—were expecting it to be telling us. One consequence of this is that we are never simply seeing what's "really there," stripped bare of our own anticipations or insulated from our own past experiences. Instead, all human experience is part phantom—the product of deep-set predictions. We can no more experience the world "prediction and expectation free" than we could surf without a wave.

When I stood backstage at the New Scientist Live festival, the stress of waiting to give my presentation sent my prediction machinery into overdrive. Given my lifetime of experience, I would not expect the floor to suddenly turn to jelly underneath me, or an anvil fall cartoon-like on my head. But my phone does vibrate in my pocket annoyingly often, causing my brain to form a kind of baseline prediction of frequent vibrations. Stress and caffeine (I had plenty of both) tend to amplify such effects, and signals from an anxious gut feed directly into the prediction machinery in our heads. When all those factors came together, that baseline prediction of pocket-buzzing briefly became my reality. But just as quickly as it occurred, I was able to reorient myself toward the facts, and recognize it as an illusion.

The illusion occurred because predictive brains are guessing machines, proactively anticipating signals from the body and the surrounding world. That guessing is only as good as the assumptions it makes, and even a well-informed best guess will frequently miss the mark. After all, there was no phone in my pocket. When the brain's best guessing misses the mark, the mismatch with the actual sensory signal carries crucial new information. That information (prediction error) can be used to try again—to make a better guess at how things really are. But experience still reflects the brain's current best guessing. It is just that each new round of guessing is a little bit better informed.

This challenges a once traditional picture of perception. Whereas sensory information was often considered to be the starting point of experience, the emerging science of the predictive brain suggests a rather different role. Now, the current sensory signal is used to refine and correct the process of informed guessing (the attempts at prediction) already taking place. It is now the predictions that do much of the heavy lifting. According to this new picture, experience—of the world, ourselves, and even our own bodies—is never a simple reflection of external or internal facts. Instead, all human experience arises at the meeting point of informed predictions and sensory stimulations.

This is a profound change in our understanding of the mind that fundamentally alters how we should think about perception and the construction of human reality. For much of human history, scientists and philosophers saw perception as a process that worked mostly "from the outside in," as light, sound, touch, and chemical odors activate receptors in eyes, ears, nose, and skin, progressively being refined into a richer picture of the wider world. Even well into the twenty-first century, leading models in both neuroscience and artificial intelligence retained core elements of that view.

The new science of predictive processing flips that traditional story on its head. Perception is now heavily shaped from the opposite direction, as predictions formed deep in the brain reach down to alter responses all the way down to areas closer to the skin, eyes, nose, and ears—the sensory organs that take in signals from the outside world. Incoming sensory signals help correct errors in prediction, but the predictions are in the driver's seat now. This means that what we perceive today is deeply rooted in what we experienced yesterday, and all the days before that. Every aspect of our daily experience comes to us filtered by hidden webs of prediction—the brain's best expectations rooted in our own past histories.

To see just how important this could be, imagine a world in which the weather forecast played a significant role in causing—not simply predicting—the weather itself. In that strange world, a confident forecast of rain helps bring about changes to the flows of matter and energy that determine the changing weather. There, a confident forecast of rain has causal powers that make *rain itself* a little more likely. There, as here, the weather forecast depends on a model (never perfect) of the way existing weather conditions are most likely to change and evolve over time. But in that world the weather you get (here and now) reflects a kind of combination of the effects of the prior forecast itself and preexisting conditions out in the world.

We do not live in that bizarre world. The weather we get is not affected by our best model-based predictions of that weather. But our mental world shares something of that remarkable profile. When the brain strongly predicts a certain sight, a sound, or a feeling, that prediction plays a role in shaping what we seem to see, hear, or feel.

Emotion, mood, and even planning are all based in predictions too. Depression, anxiety, and fatigue all reflect alterations to the hidden predictions that shape our experience. Alter those predictions (for example, by "reframing" a situation using different words) and our experience itself alters.

Consider the prickly rush of adrenaline I felt just before going onstage to deliver that speech. I had practiced attending to that prickly feeling while verbally reframing it not as a portent of failure but as a sign of my own chemical readiness to deliver a good performance. This helps alter my self-predictions, leading to a more relaxed and fluent performance. We'll explore several such interventions, stressing both their surprising scope and their undoubted limits.

What is your relationship to the reality you perceive? In what ways do you shape it, and, by extension, in what ways do you shape yourself, often without even knowing it? In this book, I draw on paradigm-shifting research to confront these crucial questions and ask what these insights mean for neuroscience, psychology, psychiatry, medicine, and how we live our lives. We'll look hard at experiences of the body and self, from chronic pain to psychosis, and see how work on the predictive brain helps explain a wide spectrum of human behaviors and neurodiversity. We'll reassess our own experiences of the world, from social anxiety and emotional feedback loops to the many forms of bias that can creep into our judgments. We'll also explore some ways that predictive brains might support "extended minds," blurring the boundaries between ourselves and our best-fitted tools and environments.

The book ends by putting some key insights into action, looking at ways to "hack" the predictive mind by changing our practices, reframing our experiences using different kinds of language, and the controlled use of psychedelic drugs. As these themes converge, we glimpse the shape of a new and more deeply unified science of the mind—one that does justice to the range and diversity of human experience, and that has real implications for how to think about ourselves and improve our lives.

THE
EXPERIENCE MACHINE

1

UNBOXING THE PREDICTION MACHINE

IT'S MORNING and I'm still asleep in my bedroom, a daunting pile of work perched uneasily beside the bed. Waking dozily from sleep I hear some gentle birdsong. Or at least, that's how it seems to me at first. But I soon discover that I am mistaken. I listen harder and realize that all is deathly quiet. Not even my cats' early mewling for food breaks the silence. I was hallucinating birdsong.

Fortunately, there is a simple explanation. My partner recently decided to ease the process of waking up in the morning by using a smartphone app that plays a birdsong instead of a traditional alarm. The app alarm starts off as a gentle chirping that very gradually, and very slowly, builds to something approximating a full morning chorus. Today, the alarm was not actually going off—it was far too early. Nor does the sound of real birdsong ever make it through the double-glazing. But I have become so used to waking up to the gently increasing tweeting of the alarm that my brain has started to play a trick on me. I now find that I quite often awake well in advance of the start of the actual alarm, already seeming to hear the faint onset of those prerecorded chirps.

These are genuine auditory hallucinations, caused by my new, strong expectation of waking to the subtle sound of the birds. There is probably nothing sinister about my proneness to this hallucination. It has long been known that hallucinations, both auditory and visual, can be quite easily induced by the right kind of training. But these, as well as a myriad of other intriguing phenomena, are lately falling into place as signs of something much larger—something that lies at the very heart of all human experience.

The idea (the main topic of this book) is that human brains are prediction machines. They are evolved organs that build and rebuild experiences from shifting mixtures of expectation and actual sensory evidence. According to that picture, my own unconscious predictions about what I was likely to be hearing as I awoke pulled my perceptual experience briefly in that direction, creating a short-lived hallucination that was soon corrected as more information flowed in through my senses. That new information (signifying the lack of birdsong) generated "prediction error signals" and these—on this occasion at least—were all it took to bring my experience back into line with reality. The hallucination gave way to a clear experience of a silent room. But in other cases, as we'll see, mistaken predictions can become entrenched and contact with reality (itself a complex and vexed notion) harder to achieve. Even when there are no mistakes involved, and we are seeing things "as they are," our brain's predictions are still playing a central role. Predictions and prediction errors are increasingly recognized as the core currency of the human brain, and it is in their shifting balances that all human experience takes shape.

This book is about those balances and an emerging science that turns much of what we thought we knew about perceiving our worlds upside down. According to that science, the brain is constantly trying to guess how things in the world (and our own body) are most likely to be, given what has been learned from previous encounters. Everything that I see, hear, touch, and feel—so this new science suggests—reflects hidden wells of prediction. If the expectations are sufficiently strong, or (as in early chirps of the bird alarm) the sensory evidence sufficiently subtle, I may get things wrong, in effect overwriting parts of the real sensory information with my brain's best guess at how things ought to be.

This does not mean that successful sensing is simply a form of hallucination, though the mechanisms are related to those of hallucination. We should not downplay the importance of all that rich sensory information arriving at the eyes, ears, and other senses. But it casts the process of seeing—and of perceiving more generally—in a new and different way. It casts it as a process led by our brain's own best predictions: predictions that are then checked and corrected using the sensory inputs as a guide. With the prediction machinery up and running, perception becomes a process structured not simply by incoming sensory information but by difference—the difference between the actual sensory signals and the ones the brain was expecting to encounter.

Since brains are never simply "turned on" from scratch—not even first thing in the morning when I awake—predictions and expectations are always in play, proactively structuring human experience every moment of every day. On this alternative account, the perceiving brain is never passively responding to the world. Instead, it is actively trying to hallucinate the world but checking that hallucination against the evidence coming in via the senses. In other words, the brain is constantly painting a picture, and the role of the sensory information is mostly to nudge the brushstrokes when they fail to match up with the incoming evidence.

This new understanding of the process of perceiving has real importance for our lives. It alters how we should think about the evidence of our own senses. It impacts how we should think about the way we experience our own bodily states—of pain, hunger, and other experiences such as feeling anxious or depressed. For the way our bodily states feel to us likewise reflects a complex mixture of what our brains predict and what the current bodily signals suggest. This means that we can, at times, change how we feel by changing what we (consciously or unconsciously) predict.

This does not mean we can simply "predict ourselves better," nor does it mean we can alter our own experiences of pain or hunger in any way we choose. But it does suggest some principled and perhaps unexpected wiggle room—room that, with care and training, we might turn to our advantage. Handled carefully, a better appreciation of the power of prediction could improve the way we think about our own medical symptoms and suggest new ways of understanding mental health, mental illness, and neurodiversity.

The Smart Camera Model of Seeing

The idea that the brain is basically a giant prediction machine is relatively recent. Prior to that, it was widely believed that sensory information is processed in a mostly "feedforward" manner—that is, taken from our senses and directed "forward" into the brain. To take the best-studied example, visual information (that older picture suggests) is first registered at the eyes and then processed in a step-by-step fashion deeper and deeper inside the brain, which is slowly extracting more and more abstract forms of information. Beginning with patterns of incoming light, the brain might first extract information about simple features such as lines, blobs, and edges, then assemble these into larger and more complex wholes. I'm calling this the "smart camera" account of seeing. But this was clearly no camera, but rather a very smart intelligent system. Nonetheless, as in a simple camera, the direction of influence flowed mostly inward, moving forward from the eyes into the brain. Only at some point quite late in this process would lifetime memory and world knowledge become engaged, enabling you (the perceiver) to understand how things are in your world.

Versions of the smart camera (feedforward) view have been influential in philosophy, neuroscience, and AI. Such a view is intuitive because we typically think of perception as all about the flow of information from the world to the mind. That picture can be found, for example, in Descartes's 1664 *Treatise on Man*. There, Descartes depicts perception as the complex opening and closing of networks of inner tubes imprinting an image of the world first onto the sense organs (such as the eyes) and then via a network of tiny tunnels deeper and deeper into the brain. As impressions from the outside world (and from within the body) flowed forward into the brain, they were said to be preserved in our minds much the way pushing your fingers into wax preserves information about their shape.

It was never clear how Descartes's mechanism would work. But what remained even as much more sophisticated scientific understandings emerged was the core idea of the perceiving brain as a relatively passive organ taking sensory inputs from the world and then "processing" them in a predominantly feedforward (outer to inner) fashion. That idea was pretty much standard in late-twentieth-century cognitive neuroscience. This was probably because it appeared as a governing principle of David Marr's hugely influential computer model of vision.

Marr was a towering figure, whose work in neuroscience, computer vision, and AI ranks among the most important contributions ever made to cognitive science. In Marr's depiction, visual processing starts by detecting basic ingredients in some incoming signal—an ordered array of pixels, for instance. From there, layered processing slowly builds toward a more complex understanding. For example, the next stage might look for places where pixel intensities display rapid changes from their neighbors—usually a clue to the presence of a boundary or an edge out there in the world. As processing moves forward, step by step and deeper into the brain, further patterns are detected, such as the recurring sequences that characterize stripes. Vision is here a matter of subjecting the raw signal to a series of operations, such as edge or stripe detection, that slowly reveal more and more complex patterns in the environment—the source of the incoming signal. Eventually, the complex detected patterns are brought into contact with knowledge and memory to deliver (though revealingly, this part of the puzzle was never satisfactorily solved) a kind of 3D picture of the worldly scene.

Marr's computer model (like any computer model) had the distinct virtue of specifying the key computations that might be involved in those early stages of processing, though the shape of those crucial final steps remained something of a mystery. The Marr model was for many years the standard picture not just in artificial vision but in neuroscience too. Even into the twenty-first century the visual system was primarily regarded as a mostly feedforward analyzer of incoming sensory information along the lines that Marr had described.

Notably absent from Marr's model, however, was another direction of influence—one running backward, from deep within the brain down toward the eyes and other sensory organs. The number of neuronal connections carrying signals backward in this way is estimated to exceed the number of connections carrying signals forward by a very substantial margin, in some places by as much as four to one. What is all that downward connectivity feeding information from deep in the brain to regions closer to the sensory peripheries doing? This wiring runs in the opposite direction to the wiring needed to perform the processing tasks described in Marr's early computational model, yet it reaches right down to those very regions.

Real neural wiring like that is costly to install and maintain. The brain, weighing in at about 2 percent of human body weight, is estimated to account for around 20 percent of total bodily energy consumption. It is by far our most "expensive" adaptive accessory. Yet a huge amount of that expense is now known to be devoted to establishing and maintaining an immense web of downward (and sideways) connectivity, spanning not just early visual processing but the whole of the brain. This is a puzzle. It was puzzling enough to lead the artificial intelligence pioneer Patrick Winston to comment, even as recently as 2012, that with so much information apparently flowing in the other (downward) direction, we confront "a strange architecture about which we are nearly clueless." Things look different, however, once we recognize the attractions of a bold new claim: that brains are nothing other than large-scale prediction machines.

Flipping the Flow

It now seems that the core operating principle of the perceiving brain is pretty much the opposite of the smart camera view. Instead of constantly expending large amounts of energy on processing incoming sensory signals, the bulk of what the brain does is learn and maintain a kind of model of body and world—a model that can then be used, moment by moment, to try to predict the sensory signal. These predictions help structure everything we see, hear, touch, and feel. They were at work when I heard nonexistent birdsong in the morning. They were at work when I felt phantom vibrations from a smartphone that was not even in my pocket. But they are also at work, as we'll see, when I hear actual birdsong, feel real phone vibrations, and see the various objects scattered about on my university desktop.

A predictive brain is a kind of constantly running simulation of the world around us—or at least, the world as it matters to us. Incoming sensory information is used to keep the model honest—by comparing the prediction to the sensory evidence and generating an error signal when the two don't match up. Despite the wiring costs, constant prediction brings many efficiencies, as we'll shortly see. It also—and perhaps more importantly—makes us flexible, able to adapt our responses in ways that reflect the demands of our current tasks and context. Instead of steadily extracting a rich picture of the world from a barrage of sensory clues, the rich evolving picture of the world is the starting point, and the sensory information is used to test, probe, and tweak that picture. Before new sensory signals arrive, the predictive brain is already busy painting a rich picture of how things are most likely to be.

This explains, in broad outline, the need for all that downward connectivity. It is carrying predictions from deep in the brain, pushing them toward the sensory peripheries. It also explains the huge energy outlay used simply to sustain the brain's intrinsic activity. That activity is necessary to maintain the model that issues moment-by-moment predictions. As a brain encounters new sensory information its job is to determine if there is anything in that incoming signal that looks like important "news"—unpredicted sensory information that matters to whatever it is that we are trying to see or do. There is increasing consensus that something like this is the primary way our brains process sensory information. Unpacking that hypothesis, the last ten to fifteen years has seen an explosion of work in computational and cognitive neuroscience that now makes detailed and testable sense of this, thereby solving the mystery of Winston's "strange architecture." That work goes by various names including "predictive processing," "hierarchical predictive coding," and "active inference." I'll mostly stick with "predictive processing" as a handy label for this family of theories.

According to this view, the smart camera picture of perception was a big mistake. Despite its intuitive appeal, the right way to think about perception is not (for the most part) as a process that runs primarily from the eyes and other sense organs inward. Nor is the brain ever just sitting there waiting patiently for sensory information to arrive. Instead, it is actively anticipating the sensory information, using everything it knows about patterns and objects in the world—the twittering sounds of birds (and of my partner's early morning alarm), the all-too-frequent intrusion of phone vibrations, and the organization of the various objects on my office desk. It is also making constant use of the active body, moving head, eyes, and limbs in ways that harvest new and better information. Instead of being a passive receiver and processor of sensory information, a brain like that is a tireless predictor (and, as we'll later see, a skilled and active interrogator) of its own sensory streams.

Bad Radios and Controlled Hallucinations

The contemporary picture of the predictive brain has historical roots in the nineteenth-century ideas of a German physicist and polymath named Hermann von Helmholtz. Helmholtz was the inventor of the ophthalmoscope used by opticians to examine the eye and formulated the law of conservation of energy. He was also interested in theories of perception and argued that we perceive the world only thanks to a kind of unconscious reasoning or inference in which the brain is asking itself, "Given everything I know, how must the world be for me to be receiving the pattern of signals currently present?" This is the question that perceptual systems are built to resolve.

You might not realize how common this is in our everyday life. If you listen to a familiar song on a radio with bad reception, the words and rhythms sound surprisingly clear. But try to listen to a brand-new song with that same reception quality and the sounds seem much more indistinct, the vocals hard to distinguish. In each case your brain, just as Helmholtz argued, is using what it knows to try to infer which words and sounds are the most likely cause of the somewhat patchy auditory signals currently being picked up by your ears. But the brain's guessing is much better for the familiar song—making it sound that much clearer. In fact, that guessing is altering the brain's responses all the way "down" to early auditory processing areas, so as to bring those responses more into line with the expected sounds. In a very real sense, your brain is now playing much of the song for itself, so the poor incoming signal is cleaned up using stored knowledge about the world.

This is the brain doing what it does best, churning out "good hallucinations" by filling in and fleshing out the missing signal according to what it expects to hear. Our brain knows about the way the song sounds and the various subtleties of that specific singer's rendition, and it can use all that prior knowledge to actively predict the most likely shape of the auditory signals as the song plays. If the world doesn't send strong counterevidence, those predictions sculpt experience, making the song sound clearer to you.

It's important to emphasize that this is not a trick of memory, so much as a fascinating window on the way perception itself works. The brain's predictions for the familiar song help it carve out the signal from the noise, rendering the sounds more clearly than the bad signal would otherwise allow. Perception of this kind is highly active. It involves sending complex predictions down the chain from higher processing areas toward the sensory peripheries, generating error messages whenever a serious mismatch is detected. This backward flow is sometimes referred to by cognitive scientists as the "top-down" flow of information. While all this goes on, the human perceiver is also active, trying to gather key pieces of sensory information by means of bodily action such as turning the head or moving the eyes. These actions too are chosen and launched by the predictive machinery, creating a unified web of mental and bodily activity. We'll have lots more to say about the role of action as our story proceeds.

Putting predictions in the driver's seat in this way makes ordinary perception into what has sometimes been colorfully described as a "controlled hallucination"—the brain is guessing at how the world is by using sensory evidence mostly as a way to correct and finesse the guessing. When inner guessing completely rules the roost, we are just hallucinating, full stop. But when it is appropriately sensitive to sensory stimulations—via prediction error signals—the guessing is controlled, and the world becomes known to the mind. When we heard that familiar song on the bad radio receiver, we were benefiting from just this kind of "good hallucination." The phantom phone vibrations we met in the Preface, though in that case misleading, were generated in just the same kind of way. All human experience, if predictive processing is on track, is built in this way. We see the world by predicting the world. But where prediction errors ensue, the brain must predict again.

The Frugal Brain

Making perception turn on prediction has another important benefit too. It enables the brain to process incoming sensory information in a way that is quite remarkably efficient. The discipline that most famously examines the issue of communication efficiency is information science, which has played a major role in developing very frugal ways to transmit signals. In the mid-twentieth century, global telecommunications systems were strained by ever-increasing demand. The problem for the telecom giants was how to convey increasingly large amounts of information using just the noisy and limited channels provided by old-fashioned telephone cables. That's where information science stepped in with a clever way to increase efficiency. The powerful technique developed by theorists working on such puzzles eventually became known as linear predictive coding, and it is still in use to this day.

Linear predictive coding has its roots in a paper published in 1948 by Claude Shannon, a mathematician and cryptographer who was working for Bell Laboratories. This crucial work showed that English text could be encoded in a very efficient, compressed way by exploiting predictability among words and letters. If a certain letter is almost always followed by another, then an efficient coding scheme can simply assume this to be so unless the case is marked as an exception. Marking only those occasional exceptions is far more efficient, using less bandwidth than would be needed to encode every letter.

Given the right expectations, even the absence of a signal can carry a large amount of information. Suppose you make a plan by telling someone that if you don't call them, then all is "as expected" and that they should therefore meet your plane at Miami airport next Wednesday at 9 a.m. local time. On the day, your failure to call amounts to a tiny one-bit signal that now conveys a very large amount of information: you will be arriving, by plane, at that time, at that place. The trick is trading intelligence and foreknowledge on the part of the receiver against the costs of encoding and transmitting all the information.

Telecommunications research has benefited from many versions of prediction-based compression. In principle, you can use anything that is already known about the most likely shape of some signal to help predict that signal at the other end, transmitting over the precious wires and cables only whatever turns out to be *different* from the predicted patterns. The receiver then updates on the basis of that residual error alone. The beauty of this procedure is that by transmitting just a few bits of error, a rich content (such as an image or message) can be reconstructed. The rich content is built mostly out of the predictions but gets anchored to reality by the residual errors.

In this way, compression by informed prediction saves on bandwidth by in effect "adding back in" all the successfully predicted elements. This is the same clever trick that enables us to economically store and transmit pictures and sounds and videos using formats such as JPEG and MP3. In the case of a picture, predictive coding works by assuming that the value of each pixel is well predicted by the value of its various neighbors. When that's true—which is in fact rather often—there is simply no need to transmit the value of that pixel. All that needs to be encoded are the deviations from what was already predicted. But that's just one simple regularity. Wherever there is detectable regularity of any kind, prediction (and hence this form of data compression) becomes possible.

Consider motion compressed coding for video. In 1959, the world was introduced to predictive interframe video coding. To get the flavor, imagine that the video is of a person running down a corridor. The only difference between frame 4 and frame 5 of the video turns out to be a slight forward motion of the runner against an unchanging visual backdrop. All that needs to be transmitted, to capture frame 5, are these few differences (the residual error) from what was already predicted. In this way you can treat frame 5 as merely a minor variant on frame 4. Perhaps the only difference lies in the positioning of the feet. Pushing that foot location information through the system is vastly less costly than transmitting a new value for every pixel in frame 5. This kind of trick is still in use today.

Now imagine a system that already knew a whole more—one that knew, for example, about all the ways different running gaits tend to continue. Such a system could use that more detailed (more "high-level") information to make predictions too, so that only unexpected foot movements would generate prediction error signals. Assuming the feet move just as expected, there's again no need to update with new information between frames. This even more intelligent system could just "hallucinate" the usual kind of ongoing motion, updating only when something unexpected occurs (perhaps the runner suddenly stumbles). However complex or high-level the predictions, it is prediction errors that must then carry the news, signaling differences from the expected and thereby keeping us in touch with a changing and sometimes surprising world.

Human brains seem to benefit from intelligent prediction strategies of just that kind, and they do so in an especially powerful way, thanks to the use of multiple "levels of processing." In these multilevel contexts, simple predictions are nested under less simple, more abstract ones—much as in our example of the running gaits, where the gait expectation is a higher-level prediction that in turn spawns predictions about the actual foot positions (a lower-level prediction). At that point, prediction errors are formed and pushed upward through the system. These nuance and refine the guessing at every level—for example, by revealing that the running gait is not the one we expected after all and selecting an alternative that does a better job.

In the predictive processing architecture of the brain, it is thought that different neuronal populations specialize in different things, so that each "higher" level can use its own specialized knowledge and resources to try to predict the states of the level immediately below it. So (to simplify) a level that specializes in predicting whole words might use its knowledge to help predict states at a lower level whose specialty is recognizing letters. But the higher level that predicts words might itself be predicted by an even higher level that specializes in whole sentences. A walkthrough example of this, along with various other nuts and bolts, can be found in the Appendix.

For now, the point to notice is that in this kind of multilevel arrangement all that flows forward (from the sensory edges ever deeper into the brain) is *news*—deviations from what is expected. This is efficient. Valuable bandwidth is not used sending well-predicted stuff upward. Who in corporate HQ wants to know that Billie's work proceeded exactly as expected? Similarly in the neural incarnation, prediction errors at every level signal only the unexpected, the stuff that may plausibly demand further thought or action.

Systems like that are wonderfully frugal in their use of the incoming stream of information. Instead of trying to deal with everything from scratch they effectively sift and filter the incoming data by highlighting only what was unexpected. This is the nugget of truth in the notion that human brains hallucinate reality. It means that the world we experience is to some degree the world we predict. Perception itself, far from being a simple window onto the world, is permeated from the get-go by our own predictions and expectations. It is permeated not simply in the sense that our own ideas and biases impact how we later judge things to be, but in some deeper, more primal, sense. The perceptual process, the very machinery that keeps us in contact with the world, is itself fueled by a rich seam of prediction and expectation.

In the next few pages, I'll try to give you some firsthand experience of the power of predictions to alter what you see and hear.

The Power of Prediction

Fig. 1.1 provides a simple illustration. Read it from the top down and then from left to right. Notice that the shape of the middle character is the same in each sequence. But the visual experience seems subtly different according to which sequence you are reading. When read top to bottom, your brain starts to expect a number there (13), while left–right

12
ABC
14

Fig. 1.1 A number/letter grid

fires up the expectation of a letter (B). These differing predictions impact the visual experience itself.

It has recently been shown that even unconscious ("masked") presentations of number/letter cues can bias us to see the ambiguous middle form in a certain way. Masking is a technique in which a stimulus is briefly shown then rapidly followed by another, different stimulus. This procedure blocks conscious awareness of the first briefly shown stimulus. Nonetheless, masked items can still impact behavior and response. In the case at hand, masked presentations of the flanking letters A and C biased subjects to classify the ambiguous central form as the letter B, while masked presentations of the flanking numbers 12 and 14 biased subjects to classify the central form as the number 13. This shows that active but nonconscious predictions bias response and judgment too. This will be important when we later apply the lessons of this chapter to more complex examples from psychiatry and medicine.

Next, consider the Hollow Mask illusion. There, a rigid mask (of the kind you might buy in a joke shop) is viewed from the wrong side, so that you are looking at the concave impression of the face. When lit from behind and viewed from a few feet away the visual experience is nonetheless one of a normal convex face—one with the nose and features clearly protruding outward. This is because we are so used to seeing faces (and so unused to seeing their inverse impressions) that the brain seems to discount incoming sensory information specifying concavity, and instead allows its own deep-set predictions of convexity to dominate. The Hollow Mask illusion is strongest for famous or very familiar faces (where we have the strongest and most detailed predictions) and is greatly weakened or abolished if the mask is turned upside down—presumably because this enables us to view it as a standard object rather than one about which we have such potent and deeply wired expectations of convexity.

To round off the parade of visual exemplars, take a look at the image in Fig. 1.2.

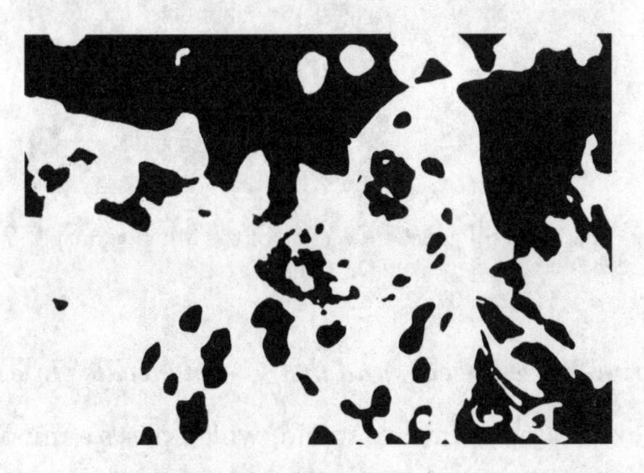

Fig. 1.2 A Mooney image

This so-called Mooney image will probably not look like anything much to you, other than a few contours and blobs of black and white. But now turn the page and look at the original grayscale image (Fig. 1.3), and then turn back to the Mooney. Your experience has been fundamentally—and probably permanently—altered. The Mooney image will now appear structured and meaningful. This impacts your actions too, as your eyes will now inspect the Mooney image in ways that track its most salient features, alighting especially upon the kitten's eyes and paws. This is a taste of one of the core effects highlighted throughout this book. The picture looks different the second time around because improved knowledge about the world (in this case, about the original picture) is enabling your brain to make better predictions.

Fig. 1.3 Full grayscale version of the Mooney image shown in Fig. 1.2

Sine-Wave Speech, and the "Green Needle" Effect

Our next examples involve sound, which poses a minor challenge given the present medium. If you are not currently near a web-enabled device, just reading the text below will suffice. But it is well worth accessing some audio when you can, as nothing beats experiencing these very simple but dramatic effects firsthand.

The first phenomenon in question is called "sine-wave speech." A sine-wave speech recording is a recording of normal speech that has been artificially degraded in a way that replaces key parts of the sound stream with pure tone whistles. It was invented back in the early 1970s at Haskins Labs in the United States as a means of studying the nature of speech perception—specifically, to test various theories about what parts of the sound stream are essential for hearing speech. Sine-wave speech sounds like a series of initially unintelligible ascending and descending beeps and whistles.

There are many demos on the web (some of my favorites can be found on the site of my University of Sussex colleague Chris Darwin, or just search for "sine-wave speech") and they all work pretty much like this. You first hear a segment of the degraded speech, which will probably make no sense to you at all. You are then played the original recording, which will consist of someone speaking a simple sentence such as "the kettle boiled quickly." After that, the sine-wave speech version with its beeps and whistles is played to you once more. The second time around, your experience is dramatically altered. Now, you clearly hear the words being uttered, and the spaces between them. It's like the experience with that Mooney image, only this time in sound.

If you practice this a few times, using different examples, you will quickly become fluent and won't even have to hear the original sentence first. Hearing ordinary speech in your native language involves the very same trick—the better your predictions (perhaps because you know the speaker or share the accent) the clearer the sounds. In every case, perception is improved by the presence of a good predictive model. Such a model uses predictions to help sort out the signal from the noise. What previously sounded like a series of meaningless beeps and whistles is now heard as a structured sentence, albeit one delivered in a somewhat distorted voice. The difference from the original experience could not be more dramatic. It is for all the world as if the original sound file has been replaced by another, very different, one.

Another example that may be familiar to some is the meme in which a repeated sound is played while one of two different phrases appears on-screen—for example, "brainstorm" or "green needle." Remarkably, what you seem to hear depends on which one you are currently looking at. Yet the difference in the way they sound is striking. This is because reading the word tips the evidential balance in favor of one auditory prediction over the other. The striking difference in what you experience again reveals the extent to which what you hear is a construct formed in part by your own predictions. Typing the words "green needle" and "brainstorm" into a search engine will enable you to try this out for yourself.

These phenomena may seem strange at first, but they are actually representative of what happens in normal perception. Every time we make sense of our worlds through perceptual encounters, we do so by means of both the incoming sensory signals *and* a rich invisible stream of knowledge-based predictions.

Hallucinating a White Christmas

Think of a song you know quite well. Now ask yourself—could I spot a sample of that song if it was extremely faint, and hidden somewhere within a three-minute sound file comprising mostly white noise? You probably don't know—it would depend, you'd think, simply on how well it was hidden. Back in 2001, researchers at Maastricht University in the Netherlands gave this task to a set of undergraduate students, who were asked to press a button if at any point they believed they were hearing the song. The song (Bing Crosby singing "White Christmas") was playing as they entered the lab, and at that time they were asked to confirm that it was indeed a familiar tune. They were then told:

> The "White Christmas" song you just heard might be embedded in the white noise below the auditory threshold. If you think or believe that you hear the song clearly, please press the button in front of you. Of course, you may

press the button several times if you think that you heard several fragments of the song.

The recording was played, and students duly punched the button whenever they felt they could detect the hidden song. After each trial, they were also asked to report on their confidence. If they were 100 percent sure they had at some point heard the song, they would rate their confidence 100, and so on down to zero, meaning they were sure they hadn't heard it at all. The trick, though, was that nowhere in the tape was there any hint of the supposedly hidden song—the tape was 100 percent white noise, and 0 percent "White Christmas." About a third of the students in that study pressed the button at least once—a significant result. The experiment was successfully replicated a few years later with an even larger cohort and it now features in a varied lineage of such experiments, all using the old chestnut "White Christmas." By manipulating our expectations (making the subjects expect to hear the faint sample of the song), the experimenters had reliably caused their subjects to experience an auditory hallucination of Bing Crosby faintly crooning away.

There are several theories vying to account for the "White Christmas" effect. For example, in the 2001 paper it was noted that those who show the effect most strongly tended also to score higher on psychological tests for "fantasy proneness." In 2011, another study found that the effect (like phantom phone vibrations) was greatly increased by both stress and caffeine. There is also some evidence that both false song detection and confidence in doing so is greater in people with schizophrenia, an area of research we will return to later. What seems indisputable is that experimentally induced expectations of hearing the song are playing a leading role in the construction of the illusory experience—just like my own predictions of hearing that birdsong alarm.

That Dress, and Other Illusions

In February 2015, a social media spark lit an unstoppable fire that spread through the internet, spawning 10 million rapid retweets and enlivening many a family dinner conversation. The spark was a picture of a dress, due to be worn to a wedding in Scotland. As almost everyone reading these words will recall, many viewers saw that dress as clearly gold and white while others were utterly convinced it was blue and black. If you were away on Mars that year, check it out online. I belonged firmly to the gold and white camp. But we gold-and-whiters lose (at least insofar as it is possible to lose here at all) as the actual dress, when viewed under normal lighting conditions, will appear blue and black. How do we make sense of this radical variation between experiences?

Fig. 1.4 shows a version of the so-called Ponzo illusion. The two heads are exactly the same size (go on, measure them). But in the real world, the best explanation (given the perspective) would be that the rear head is something of a giant. This shows—just as Helmholtz thought—that what we see is not

Fig. 1.4 A version of the Ponzo illusion.
The two heads are exactly the same size on the page.

simply how things are: rather, we see whatever our brain infers (guesses) as the most likely cause of the evidence coming in from the senses.

Much the same reasoning applies in the case of the dress. But there, disagreement occurs because different people's brains seem to be assuming rather different things about the depicted scene—in particular, the opposing (blue versus gold) camps make different assumptions about the lighting in the room. These include assumptions about the general level of brightness in the room, the positioning of the light source, and whether the dress is in shadow or not in shadow. Brains that make different assumptions about those conditions will make different inferences about the color of the dress, leading some to see the dress as clearly and indisputably blue, and others as clearly and indisputably gold.

Confirming this theory, Pascal Wallisch and a team based at New York University conducted an online survey of 13,000 subjects, who were asked not just about how they saw the dress in the photo, but also about how they believed the lighting to be for the photo—did they think it was shot in artificial light, natural daylight, or were they unsure? Sure enough, there was a strong correlation such that those who said that they assumed natural light tended to see the dress as white and gold, while those who thought the lighting was artificial tended to see blue and black, with those who were unsure displaying a more varied mixture of responses.

Why should different individuals make such different assumptions, given that we share a common world? Here, there's an intriguing twist. The respondents were also asked whether or not they self-identified as "larks" or "owls." Larks are those who tend to get up early, go to bed early, and feel best in the morning, while owls have the opposite profile, preferring to sleep in, staying up later, and feeling best at night. Remarkably, these self-identified "circadian profiles" correlated strongly with how the dress was perceived. Larks tended to see the dress as white and gold, owls as blue and black. The authors of the study conjecture that those whose daily routines tended to deliver more experiences of one kind than the other (natural versus artificial lighting) approached the photo differently as a result, making lighting assumptions according to their own history of perceptual encounters with the world.

In one way, this is unsurprising. The brain's predictions about the nature and origins of a light source have to be based on something, and individual history must matter in that regard. In another sense it is quite astounding to think that such a stark difference in how we see a simple photo can be rooted in—indeed, quite delicately responsive to—our own idiosyncratic daily habits. This is our first encounter with something that will loom large in our discussions—the impact of our own daily actions upon our brain's predictive models. Our own actions and histories sculpt the onboard prediction machinery that in turn sculpts human awareness, right down to the level of what seem to us to be basic sensory experiences—such as my seeing that dress as unmistakably gold.

Learning to Predict

Predictions help structure all our experience, so the question naturally arises—where did those predictions come from in the first place? Surely, we must already be able to perceive and experience the world in order to learn to make the predictions? How can we learn a good predictive model if perceiving the world depends upon having a good one in place already?

To some degree, we are obviously not starting from scratch. Millions of years of evolution have determined the bedrock configuration of the machinery we command at birth: the early wiring of the brain, the structure of our sense organs, and the shape of our bodies. Courtesy of all that, we start our journey already armed with plenty of hard-won knowledge. You might even say that (in a slightly strained sense) creatures with lungs are already structurally "expecting to breathe." But evolution has left a lot still to do, and creatures like us specialize in learning about their worlds on the basis of repeated sensory encounters.

It is here that the prediction machine gets to play another, deeply complementary role. For it turns out that by trying to predict our own sensory flows we can drive learning. This means that just by attempting to predict the world we can acquire the knowledge that later enables us to predict the world better. This can seem a bit like magic, conjuring a good predictive model from thin air. There's no magic involved, but the conjuring trick is nonetheless impressive! By trying (and failing) to predict the world, we can learn to do better, until our predictions succeed.

In thinking about this process, it's important to distinguish the raw sensory evidence (such as the incoming patterns of light and sound) from the meaningful perceptual experiences that we form as a result. In the absence of a good-enough predictive model, we will not succeed at turning the raw evidence into anything approximating a coherent understanding of the world. It will be like viewing those Mooney images, or worse. But even so, the brain can still manage to learn. It does so by looking for better and better ways to predict that unruly sensory barrage. Very young infants seem to spend most of their time doing just this, trying to find useful patterns in the sensory stream.

Well-understood machine learning techniques show exactly how this is possible. By trying again and again to predict the stream of sensory evidence, certain kinds of systems can slowly improve their initially awful performance until a useful predictive model has been built. Such systems can even start by using a randomly generated "model" whose predictions, unsurprisingly, are very bad indeed. But every time the artificial neural network fails to generate a good prediction, it alters its own processing routines, making it just a little more likely to do better the next time around. Over time, such a process unearths methods of making good predictions. In this way a good predictive model can be learned by starting with a very bad (or totally random) one, and then slowly following a gentle gradient of improvement.

The great thing about learning to predict by trying to predict is that the world itself is constantly correcting your failures. If I wrongly predict the next word you are about to utter, the next thing that hits my ears is a sound stream corresponding to the correct word. My brain can use that information to try to improve its predictions next time around. This is intuitive. For example, one way to predict quite a lot about the most likely next word in a sentence is to implicitly know a lot about grammar. But a good way to learn about grammar is to try, again and again, to predict the next words you are going to hear. As those attempts continue, your brain can slowly, unconsciously, discover the regularities that will enable you to do a better job.

Perceiving as Predicting

So how does your predictive brain work when you're out in the real world, away from Mooney images and sine-wave speech? Let's say you are out camping in the forest, and all around you is the quiet, peaceful bounty of nature. You've been huddled in your tent all night, and when you step out early the next morning, your friend is already awake and quickly gets your attention. They point above your head and say, "look at those trees over there." What steps is your mind taking as your eyes follow to where your friend is pointing? Remember that our brains are never starting from scratch—even when you first wake up in the morning in a new hotel or on vacation, your brain is already busy predicting something. As you look up, new waves of sensory information arrive. There is reflected light hitting your retina. There may also be sounds reaching your ears, smells reaching your nostrils, and various tactile stimulations of your skin. Plus (and we will have lots to say about this later in the book) many internal signals coming from your gut, heart, and elsewhere.

Sticking for simplicity's sake with just the reflected light, this stimulates cells in your retina and they send signals upstream. As those signals reach early visual processing areas, they are compared to the signals your brain currently expects. Perhaps your brain expects only some rather nondescript trees. This may be so if, like me, you are not an expert forester. Or it could be predicting something much more specific. Perhaps you knew you were in a certain part of the forest and your brain was predicting visual information of a very specific kind—the kind you'd be getting if the trees in front of you were mountain ash.

Either way, you were not predicting the robin you now see perched on top of the tree. As your brain's best predictions confront the sensory evidence, it is these residual differences that matter. These cause "prediction error signals" that encode the sensory information that your brain didn't so far manage to predict. These error signals flow forward (and sideways too), pushing deeper into the brain where they are used to generate new, improved attempts at guessing. The error signals push, pull, and probe to see if there is stuff that you already know that would generate a more successful prediction, one better able to match the actual sensory signal. As better downward-flowing predictions emerge, so does finer detail, including more about the look of the trees, and that little robin.

That robin might be a total surprise to you—perhaps they are not usually around at this time of year. But predictions and prediction errors are exchanged very rapidly and so we are not aware of all the frantic activity going on just beneath the surface. As far as you are concerned, all that happened is that you looked where your friend was pointing and simply saw the tree, complete with a (somewhat surprising) robin perched happily on top. Of course, you don't first see a rough tree outline, then a better one, this time with a bird added. It's all happening too fast for that. But there's a sense in which that's exactly what is going on inside the brain.

Fig. 1.5 provides a useful illustration of this process. The top image depicts the traditional view of how the brain processes information. The part on the left represents the raw sensory data hitting our retinas, and the overlaying cartoon in the panel to the right represents the process of starting to extract information about what's out there from that data. But in the predictive processing view beneath, we begin not with the raw data but with a prediction—in this case, of a rather generic tree. As incoming sensory data is compared to that prediction it soon emerges that there is something unexpected here, something that the generic tree prediction was not even close to accounting for. Prediction error signals result, initiating a back-and-forth process in which revised predictions meet the sensory evidence. After a flurry of such exchanges the brain settles into a stable interpretation of the scene—one that now includes the unexpected robin, and one that will also (though this is not depicted here) flesh out more details about the tree.

A word of caution. It's important not to think of what I will often be calling the "sensory evidence" or the "raw sensory signal" as itself something that is experienced. Instead, experience is what happens as sensory evidence (for example, patterns of reflected light impacting cells in the retina) gets matched by better and better predictions of that evidence. These predictions are the distilled fruits of previous experience and learning. The initial predictions act like a rough

TRADITIONAL VIEW

PREDICTIVE PROCESSING VIEW

Fig. 1.5 On the traditional view (top), sensory information is collected and passed up the chain, where it is matched to memories and activates more abstract understandings. On the predictive processing view (bottom), you start with an active model (your brain's best guess at what's likely to be out there) and use resulting prediction errors (here, from the unexpected visual information specifying the robin) to refine and correct the guessing.

draft of experience. Our hard-won baseline knowledge about the world usually makes for a good first attempt. But experience is constituted by the rapid rewriting (on the basis of resulting prediction errors) of those drafts. In other words, experience reflects the way our initial predictions are adjusted as prediction errors, flagging sensory information that wasn't yet predicted, flow around the system. Those errors flag the unexpected and demand new and better predictions. Fig. 1.6 depicts this flow. For a fuller exposition, see the Appendix.

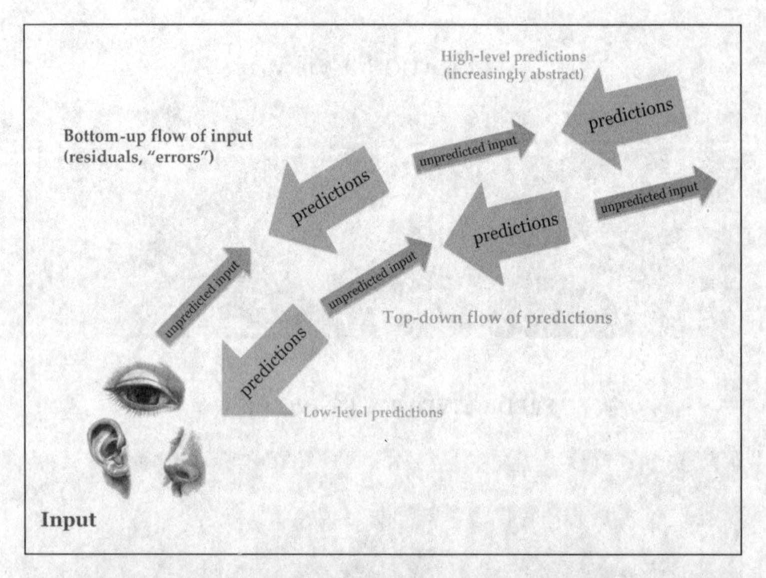

Fig. 1.6 A highly schematized look at the predictive processing view of information flow in the brain. Sensory inputs are processed in the context of predictions (based on prior knowledge and experience) originating from deeper inside the brain. Prediction errors flag unpredicted parts of the sensory signal. The errors flow forward, recruiting revised predictions.

•

From phantom phone vibrations, to hallucinations of birdsong and "White Christmas," to seeing an actual robin in the forest, our experience is shaped through and through by the brain's ongoing attempts at prediction. Brains are prediction machines, and the way we experience both the outside world and even our own bodies reflects this simple but transformative fact. This changes how we should think about our own minds, the evidence of our senses, our bodily feelings, our medical symptoms, and perhaps ultimately our contact with reality itself.

2

PSYCHIATRY AND NEUROLOGY: CLOSING THE GAP

"PAIN DON'T hurt"—so claimed Patrick Swayze's character Dalton in the 1989 movie *Road House*. But we all know pain does hurt. My partner and I own an old houseboat moored on the Calder and Hebble Navigation in the beautiful English countryside of West Yorkshire. Cruising down that canal can be a bit like threading your way through a jungle waterway rich with lush, dense, overhanging foliage. Atmospheric for sure, but that lushness is supported by some serious rainfall and all too often the steps leading down to the back door of the boat become slippery. One exceptionally rainy day my partner—who was once a medical doctor and is now a neuroscientist—fell while descending the steps. She lay in extreme agony on the hard, wet metal stairs, fearing a broken back while we waited impatiently for the paramedics to find their way to the boat. This proved to be no easy task on a featureless stretch of canal. When the paramedics finally arrived, they quickly determined that the injury (though bad) was not that extreme. But even once that alarming moment had passed, my partner still couldn't move off the steps until some serious meds had kicked in.

But now consider another accident. A report in the *British Medical Journal* describes the case of a construction worker who had jumped off some scaffolding. Beneath him, to his horror, was a 15 cm nail that pierced clean through his boot when he landed. The man—like my partner—was in agony, tortured by every small movement of his foot. He was given some even more powerful sedatives, fentanyl and midazolam. But when doctors removed the boot they discovered that the nail had not penetrated his foot at all. In fact, it had passed safely between his toes. There was no bodily injury causing the excruciating pain he felt, though it was completely genuine. In his case, however, the experience was produced entirely by his own powerful prediction machinery. Those searing pains were false perceptions created by his brain's predictions (based on the visual evidence) of serious injury and the kinds of feelings that might result.

What this case and many others in this chapter show is that pain can sometimes be remarkably disconnected from standard bodily causes. Such disconnections, and Dalton's bald assertion that "pain don't hurt," become much less puzzling once we realize that brains build human experience only by combining their own predictions with sensory evidence. In the same way that chronic expectations of incoming calls had caused me to feel phantom phone vibrations in my pocket, strong expectations of pain (from seeing the ripped boot and the protruding nail) caused the construction worker to experience agonizing pain. As we'll later see, the same kind of thing can occur over longer periods of time and in less obvious ways too, allowing not just pain but many other medical symptoms to be genuinely experienced despite the absence of the usual kinds of physiological cause. Such effects seem much less surprising once we accept that what we might think of as simple or "raw" sensory evidence is itself never experienced. Instead, experience always and everywhere reflects those rich webs of prior knowledge and here-and-now expectation.

When we see a large red beetle crawling along a branch,

we are not seeing the responses of the photoreceptors in our own eyes. Their activity is simply one of the sources of evidence that leads the brain (given the rest of what it knows) to infer the presence of such a beetle. Similarly, responses of the "pain receptors" (known as nociceptors) are not what we feel when we're gripped by a sharp pain. Instead, those responses are simply one source of evidence, acting in concert with a rich background of knowledge and belief. That's why we can genuinely feel pain even when nociceptor activity is absent.

We can also fail to feel pain even when intense nociceptor activity is present, perhaps because we're too busy acting in order to survive. What we feel is in every case a construct (just as Helmholtz suggested). It is a construct that reflects a process of unconscious inference—informed guessing—about the nature of the events causing our sensory stimulations. Sometimes the result of that informed guessing is an outward-looking experience, for example "seeing a red beetle," while at other times it is an inward-looking experience, such as "severe pain in my left foot." But the process is essentially the same.

Predictive processing provides a deep, unifying framework within which to make sense of all these effects. In so doing, it opens the door to new ways of thinking about human experience, and the many ways it can vary. Understanding the nature of pain matters. But it is really just a useful starting point for that much larger project of explaining the full diversity of human experience. Appreciating the role of brain-based predictions in the construction of all our experiences pays conceptual dividends too. It invites us to see beyond old and unhelpful dualisms such as "mental" versus "physical" and "psychiatric" versus "neurological." This may be the most important legacy of a better understanding of the predictive brain in action.

Beyond Tissue Damage

In its most basic manifestation, pain alerts us to actual or imminent bodily injury. This is clearly a crucial and highly adaptive function—an indicator that there is a problem that needs to be addressed. But when pain arises without trauma, or (as in cases of chronic pain) persists long after the normal healing period, it becomes a problem in its own right. It is estimated that up to 10 percent of the world's population experiences chronic pain. In the U.K. alone, a 2016 meta-analysis suggests that between one third and one half of the population experiences chronic pain. It is a major burden both on health care systems and the global economy.

What causes chronic pain? The notion that pain is a simple, direct response to bodily damage has long been abandoned in both clinical practice and the sciences of mind. Instead, pain has generally been assumed to fall into two different categories—nociceptive and neuropathic. Nociceptive pain is pain that is performing its adaptive function, indicating actual or threatened bodily damage—as when you feel the stab of a cut finger, the piercing agony of a broken bone, or the throbbing pain accompanying an infection. Neuropathic pain, by contrast, is defined as pain that is caused by damage or disease affecting the sensory systems that deliver the experience of pain, or the processing of pain information. Nociceptive pain is telling us that something is wrong in the body. But neuropathic pain (such as diabetic neuropathy, where limb pain results from nerve damage caused by high blood sugar) is more like a malfunction in the pain-signaling system itself.

Nociceptive pain can be compared to the warning light in a car when it correctly indicates some kind of mechanical or electrical problem. Neuropathic pain is more like a faulty warning light—a constant intrusive signal caused by damage to the warning light wiring. But as scientists delved deeper and deeper into the nature and origins of pain, even these two very broad categories struggled to accommodate all the cases. In 2016 a third category was added, known as "nociplastic pain." This was defined as pain arising from abnormal processing of pain signals without any clear evidence of either tissue damage or any other recognized systemic pathology. In other words, the warning light is on but there is simply no obvious cause—not even damage to the warning light wiring itself.

Predictive processing offers some tantalizing clues about this final mysterious category. A repeated theme in a burgeoning literature is that conscious and nonconscious expectations about our own states of pain can make a surprisingly large difference to the amounts of pain we experience. In work dating back to the 1990s, Professor Irene Tracey and colleagues at the University of Oxford showed that expectations of pain activated key neural circuits relating to the experience of painfulness. In one striking fMRI study, they showed that religious beliefs could regulate the experience of physical suffering, arguing that a kind of high-level reframing of the sensory signals mediated their actual experience and exerted an analgesic effect. When shown religious images, religious subjects rated a sharp pain as less intense than atheists shown the same image. But alter the image to one lacking such significance and the pain ratings were equal for both groups. The potential for various forms of active reframing to alter pain experience is a fascinating topic that we will return to (in Chapter 7) as just one among many ways to "hack" our own predictive brains.

Recent work suggests that many such effects depend both on what we (consciously or otherwise) predict and, as we'll shortly see, the way we attend to our own bodies. Just as in the case of outward-looking sensing using eyes and ears, the brain does not passively wait for inward-looking pain information to arrive via the nerves. Instead, it proactively predicts the arrival and intensity of pain information and estimates the likely reliability of its own predictions, up- or down-regulating experiential pain accordingly. Even simple verbal manipulations such as the dentist describing the feeling you are about to experience as a "gentle tickle" alter the way you experience the effects of the drill. But those effects depend not just on the words you hear but also on your level of confidence in what the dentist tells you.

To see why, we need to add a final—and crucial—piece to the predictive processing puzzle. We have seen that human experience arises at the meeting point of predictions and sensory evidence. But exactly how those two potent forces meet and balance is flexibly determined by a further factor: the brain's best estimate of their relative reliability and significance. Predictive processing accounts refer to this as their estimated "precision" and incorporate it as a varying weighting on predictions and on sensory stimulations. According to these accounts predictive brains are constantly estimating precision and changing the way they treat sensory evidence and their own predictions accordingly. This means we need to think not just about our brain's predictions and the incoming sensory evidence, but also about the way these estimates of precision flexibly alter the balances of power between them.

Precision is thought to be estimated in neuronal populations throughout the brain. It is the resulting "precision variations" that offer room for maneuver on the long road leading from raw sensory stimulation to human experience. For Dalton in *Road House*, such variations might allow him to exert a degree of control over his own experiences of pain. In the case of the construction worker, unwilled precision variations caused him to experience incapacitating pain despite the absence of physical injury. His pain was constructed in part as a response to visual information (seeing the nail through the boot) that seemed to offer strong, reliable evidence of serious bodily harm.

Multiple studies show the impact of estimated reliability (precision) on experiences of pain. In one such study, experimenters used heat stimuli to induce different pain intensities while manipulating the subjects' expectations about its likely magnitude. The experimenters created confident expectations in the subjects by truthfully telling them when they were about to receive a low-, medium-, or high-intensity heat stimuli, or telling them to expect an "unknown" level. How did their confident expectations alter their perceptions? When the subjects had reliable expectations of intense pain, high-intensity stimuli were perceived as being extra painful. Similarly, low-intensity stimuli were experienced as even less painful when they were presaged by the low-pain verbal cue—rather like the dentist's "gentle tickle." But all these effects disappeared when predictions were rendered uncertain. This result fits well with the idea that the brain's best estimates of precision (reliability) play an integral role in shaping our experience. Only predictions that our brains estimate to be reliable get to exert a powerful influence on our sensations. If you really don't trust your dentist, then all bets are off.

However, it has also been shown that prediction-based effects on pain intensity can sometimes be induced without engaging conscious expectations at all. Using standard techniques (such as rapidly flashing a visual cue so that it is registered only subliminally), it is possible to create strong (precise) nonconscious predictions of imminent pain. In one experiment, different subliminally presented visual stimuli (pictures of two different male faces) were paired with differing intensities of subsequent electric shock. Once this face/pain association was unconsciously learned, a shock administered following the subliminal presentation of the "low-intensity" face was experienced as less painful than when the same shock was delivered following the high-intensity cue (the other face). This pain reduction effect remained in place even if the same face was later shown long enough to enter conscious awareness. Despite never consciously experiencing the different face/shock associations during the learning phase, the brain's prediction machinery had clearly picked it up, and was using it to sculpt our experience of pain.

This confirms that conscious predictions, confidence, and expectations form just a small part of the complex multilevel prediction machinery that delivers human experience. They are just the tip of the predictive iceberg.

Placebo and Nocebo Effects

It is widely appreciated that symptomatic relief can sometimes be obtained without the use of any clinically active ingredients—the so-called placebo effect. These effects run surprisingly deep. Placebo-induced changes have been shown to reach far down, altering responses even at the level of the spinal cord. This is good evidence that they are not simply superficial effects on high-level reporting. Instead, our active expectations of pain or relief are somehow impacting the whole web from which experience itself is constructed.

A striking example is "placebo analgesia" whereby an inert treatment causes pain relief. This is relatively easy to induce and surprisingly effective in many cases. As with the classic sugar pill treatment, what makes placebo analgesia effective is that it activates conscious or unconscious expectations of relief. The higher the patient's estimate of the power of the intervention, the greater its effect. Inert substances delivered by syringe, for instance, are typically more effective than those delivered by pill, presumably because we automatically estimate this as a more powerful form of intervention. Hypnosis can produce similar effects and is sometimes so effective that some people can comfortably undergo invasive surgery under its influence.

In a laboratory setting, some forms of hypnosis were shown to greatly increase pain threshold following dental pulp stimulation (a procedure whose very name gives me a chill) with full analgesia being obtained in 45 percent of patients and delivering an average pain threshold increase of 220 percent. Doctors will also sometimes prescribe "impure placebos"—drugs that (unlike real placebos) do have some clinically active ingredients but ones that are simply not relevant to the patient's specific symptoms. Impure placebos are often very effective, presumably because they too activate potent expectations of relief. A large 2013 survey in the U.K. found that 97 percent of family doctors had at some time prescribed a placebo (pure or impure) to a patient.

But of course, if expectations can improve outcomes, so too can they worsen them. Nocebo effects are the inverse of placebo effects and occur when expectations and predictions cause us to experience unwanted symptoms rather than relief. For example, if your doctor applies a 100 percent innocent cream but warns you that "many patients will experience intense and unpleasant itching sensations," you may experience intense itching as a direct consequence of your expectations. While it may be laudable to carefully list all known side effects of prescription medicines, and even necessary for informed consent, such warnings can sometimes actually bring about the unwelcome effects they describe. This can create another self-perpetuating cycle. For even if only a very small fraction of patients would otherwise experience these specific ill effects, the mere fact that we are told about them by our doctors (or read about them on the packaging) can itself act to increase their apparent incidence, leading to further warnings that rapidly entrench expectations of the ill effects.

Self-Confirming Cycles of Pain

Many studies have demonstrated the profound impact of expectations on pain and relief. But there are plenty more twists in store. In one important recent study, experimenters showed how expectations about pain intensities can become self-fulfilling over longer time spans too, creating another circular self-reinforcing cycle.

In the studies, participants were first shown arbitrary abstract visual cues (two different geometric shapes), each paired with an image of an analog thermometer showing either a high or low reading. Over repeated exposures to these pairings, participants learned to associate the shapes with the readings. They were then shown the geometric shapes (cues) without the associated thermometer reading, but while a pad applied to the inner forearm or lower leg administered a precisely controlled, painful level of heat. What the subjects did not know was that the intensity of that applied heat was always the same. Regardless of which geometric cue was shown, the applied heat was always about 48 degrees C (a little over 118 degrees F).

By keeping the actual intensity of the heat constant in this way, the experimenters were able to isolate the effects of subjects' learned expectations on experienced pain. While undergoing an fMRI, subjects rated how much pain they expected, and how much subjective pain they felt they then received. The brain imaging data allowed the experimenters to look not just at these reports of expected and experienced pain, but also at the underlying neural activity itself. Specifically, they were looking for a complex brain imaging signature that appears to correlate well with the experience of physical pain. This enabled the experimenters to check whether what the subjects said they were experiencing was also reflected in the kinds of neural activity that would usually indicate pain.

The results were clear. Experienced levels of pain were pulled upward or downward depending on the cue-based expectations, and these effects were reflected in the neural pain signatures. This is just what all that earlier work would have led us to expect. But—and here's the twist—these experimentally biased experiences induced expectations themselves. Seeming to experience higher or lower levels of pain in the ways cued by the geometric shapes created a feedback loop in which participants' own experiences appeared to confirm their (false) cue-based expectations. This "false confirmation" cemented their misguided belief in the predictive power of the different cues. You might have expected subjects to learn, over time, that the cues were not reliable, but this did not occur. But then again, how could it? Each time the heat was applied, the different geometric cues caused them to experience its intensity in line with the false expectations they had learned. So every exposure seemed (subjectively speaking) to simply confirm the predictive power of the cues!

Such studies suggest a complex dynamic in which false expectations, once they get a grip on us, become increasingly resistant to change. This phenomenon of spuriously self-confirming expectations is probably more common than we realize, as when a patient, expecting dentistry to hurt, experiences greater pain than they otherwise would—which then in turn appears to confirm, and thereby cements, their own prior belief.

Functional Disorders

The predictive brain hypothesis (and especially detailed work on predictive processing and active inference) provides a unifying framework through which to understand a wide range of psychological phenomena. As such, it plays a leading role in new approaches to mental illness (and mental difference).

In traditional psychiatry, diagnoses are made, and treatments are given, on the basis of loosely correlated sets of symptoms and associated chemical changes in the brain. But an emerging multidisciplinary approach known as "computational psychiatry" approaches mental health and illness in a more fundamental way. Arising at the crossroads of neuroscience and computer models of the mind, computational psychiatry aims to develop a more insightful and systematic alternative to the standard symptom-based approach. It seeks to understand psychiatric conditions (and psychological diversity more generally) as a reflection of differing balances in the ways our brains process information. Where such an understanding is possible, symptom clusters begin to make better sense, and treatment options (where treatment is appropriate) better motivated. The hope is thus to approach mental health and mental illness in the kind of principled and evidence-led way we now approach physical health—by, for example, trying to understand and manipulate the deep causes of aging and cancer, rather than simply treating the various surface symptoms as they arise.

An especially revealing range of cases, perched uneasily between the standard remits of psychiatry and neurology, involves what are now known as functional disorders. These are cases where symptoms such as motor problems, paralysis, or even blindness are present, but no standard physiological cause can be identified. Other names for this include "medically unexplained symptoms," "conversion disorders," "psychosomatic," "psychogenic," or even (in the thankfully quite distant past) "hysterical" disorders. Functional neurological disorders are entirely genuine but appear not to be caused by any kind of anatomical or structural change or conventional disease process. The term "functional" reflects the fact that some aspect of normal functionality (typically involving sensation or movement) is altered or lost despite the apparent absence of any structural or recognized neurological cause—in other words, there is impairment without evidence of systemic damage or disease. Importantly, the presence of a functional disorder is not—and should never be taken to be—evidence of faking or "feigning." Instead, the impairment or disability is real, and there is no implication that it is under deliberate control.

It may be worth pausing for a note on terminology. I will use "structural disorder" and "structural damage" to refer to any cases where there is a standard neurological condition, bodily injury, or disease process present: one whose action already accounts for the experienced pain, disability, or sensory change. I will contrast this with cases of functional pain, disability, or sensory change where similar symptoms are experienced but without any evidence of sufficient, persisting structural causes. In much of the literature, this same distinction is marked using the much more problematic terms "organic" versus "functional"—using organic to mean cases where there is some standard neurological condition, bodily injury, or disease process present. I avoid this usage because, to be blunt, it is nonsensical. Functional disorders are as "organic" in origin as any others, and that is in fact one of the most important things that a predictive processing approach can help us to appreciate.

Functional disorders can be consequences of emotional trauma or stress, but they may also appear in the aftermath of accident or injury, if the impairment inexplicably persists long after normal healing processes are complete. They are also frequently (and somewhat confusingly) interleaved with various kinds of impairment and disability whose origins are structural, such as the presence of injury or disease. Nor are they rare. They are the second most common reason (after headache) for new outpatient neurological referrals, accounting for around 16 percent of all such cases. Functional disorders can present as unexplained cases of blindness, deafness, pain, fatigue, weakness, abnormal gait, tremor, and seizures—in fact, just about any possible impairment. To further complicate matters, many real-world cases present as a puzzling mixture, where physical disease or damage is actually present but is insufficient to account for the degree or variety of pain and incapacity actually experienced. In other words, there are differences in severity of pain or impairment that seem not to be explicable by immediate reference to the underlying cause.

One perfectly proper response in such cases is, of course, to humbly note that there are often standard underlying causes that have either been missed by the physician or are currently unknown to science. But sometimes, as we'll see, the evidence points to a different kind of cause—one involving altered balances within the predictive brain. When this happens, the diagnosis of some form of functional neurological disorder remains a delicate matter. Sufferers will often resist that diagnosis, thinking that they are being told that their very real problem (pain, paralysis, tremor) is in some sense "all in their head." But as we come to a better understanding of the way all human experience, including medical symptoms with more standard physical causes, is constructed, this kind of stigma may increasingly be avoided.

What could reasonably lead medical practitioners to diagnose a functional neurological disorder? The most striking form of evidence is that the contours of functional problems often follow our intuitive notions of disease or anatomy rather than medically or physiologically sound ones. An example is "tubular" visual field defect. Here, patients with a functional loss of their central visual field often report a visual "dead zone" of exactly the same diameter, no matter how close or far away their visual field is tested (see Fig. 2.1). Such a tubular deficit pattern is straightforwardly optically impossible: any visual field deficit must seem to affect more of the visual field

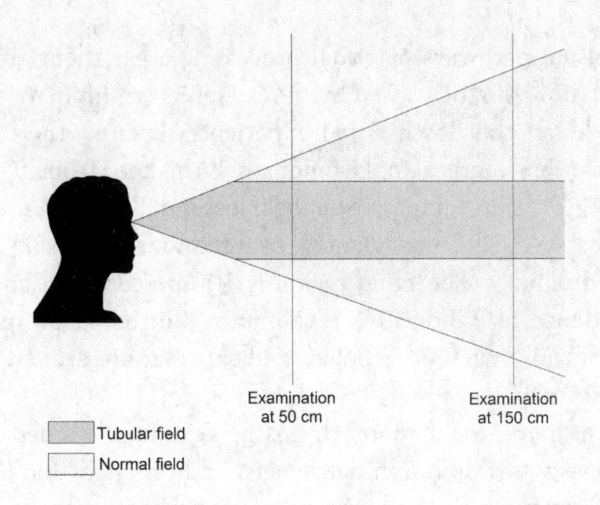

Fig. 2.1 Tubular visual field defect. Such a pattern of impairment is optically impossible.

when examined at 150 cm from the eye than it does when tested at 50 cm from the eye. For example, if you hold this book up 50 cm from your face it will appear to occupy more of your visual field than it would if viewed from 150 cm.

Tubular visual field defects do not conform to this optically inevitable pattern. This shows that the defect is functional in nature: there is a genuine loss of visual function, a blind zone, that simply cannot (given the laws of optics) reflect underlying damage to the visual system itself.

What could be going on? The important point to appreciate is not that the pattern of impairment here is optically impossible (though that is a solid clue that there is something unusual going on). Rather, it is that the shape of the impairment follows the shape of the person's own expectations and predictions. Their brain strongly predicts a uniform tunnel-shaped loss of vision and that is what they then experience. That kind of uniform-diameter loss is not optically possible and could not be caused by any form of structural damage to the visual pathways or visual processing areas themselves. It could, however, be caused by the person's own hidden predictions about their own visual experience. Even so there is no implication—and again, I cannot stress this too strongly—that the person with tubular visual field loss is deliberately predicting that loss, and thus feigning, or intentionally causing, their own disability. The blind region is 100 percent real and the experience of blindness is involuntary. But its shape reflects the way our own hidden beliefs and expectations are sculpting our experience.

Another—even more dramatic—example comes from University of Edinburgh neurologist and professor Jon Stone. Stone recounts the tale of a teenager whose vision progressively worsened until one day she woke up effectively blind. Extensive tests revealed nothing structurally amiss with her eyes or brain. She was diagnosed as having a functional neurological disorder. In the past, it was thought that such disorders were always due to abuse, stress, or trauma. Nowadays, it is clear that this is not the case. Abuse or trauma can be precipitating factors, but so can physical injury, other forms of disease, or sometimes nothing (nothing obvious at any rate) at all. In the case of the blind teenager, it turned out that she had a history of migraines and that these were triggered by light. As a result, she spent long periods in dark rooms.

Stone suspected that the teenager's aversion to light and her increasing experiences of darkness had somehow tricked her brain into constantly predicting darkness, and that this lay behind her functional blindness. To push back against those hidden predictions, he showed her that her brain was still getting good sensory evidence via her eyes. He did this by, for example, pointing out that she was often making eye contact with him or copying his gestures—despite none of that making it into her conscious awareness. He also used a technique known as TMS (transcranial magnetic stimulation), which uses a magnetic field to induce activity in neurons in the brain. Carefully applied, this causes the visual centers to fire. Under those conditions, she "saw" phosphenes (flashes of light) and this, Stone conjectures, may have helped her brain to learn that the predictions of "not seeing anything" were misguided.

With these interventions, and with a lot of careful scene setting in which Stone and colleagues explained the nature and possible origins of functional disorders, the teenager regained her sight, eventually making a full recovery. Of course, it is impossible to prove that the recovery was a direct result of the interventions. But Stone documents other such cases, in which recovery occurs following similar patterns of explanation and intervention. Predictive processing offers a compelling general picture that makes sense of both the existence of functional disorders, and the apparent efficacy of these forms of treatment. At the core of that picture is the idea that predictions about our own sensory capacities or physical abilities are playing a key role, causing genuinely experienced symptoms to fall into line with those hidden expectations.

So just how could this work?

Disordered Attention

What we need to understand is why, in these cases, misplaced predictions and expectations get to play such a strong role, effectively carving new experiences out of whole cloth. The explanation for this seems to lie with hidden disturbances to the brain's mechanisms for estimating precision—mechanisms that deliver attention in all its forms. This can seem a little technical but it is worth understanding, since aberrant precision estimations are now thought to be implicated in a wide range of psychiatric and functional disorders, as well as determining the range and variety of more neurotypical response.

Recall that precision, in these models, is simply a weighting factor that can amplify and dampen different aspects of processing. In brains, precision-weighting involves the action of (among other things) complex neurotransmitter systems centered on dopamine and other chemical messengers. Their coordinated action amplifies some aspects of neuronal activity at the expense of others. Precision variations act rather like a volume control, altering the downstream (post-synaptic) influence of whole populations of neurons. But there is not just one volume control in play but many. There are many such controls because precision is thought to be estimated at all times and for all neuronal populations. Varying estimates of precision alter patterns of post-synaptic influence and so determine what (right here, right now) to rely on and what to ignore. This is also the way brains balance the influence of sensory evidence against predictions. In other words, precision variations control which bits of what we know and what we sense will be most influential, moment by moment, in bringing about further processing and actions. Expressed like that, the intimacy of precision and attention is apparent. Precision variation is what attention (a useful but somewhat nebulous concept) really is.

For example, suppose I want to find a needle recently dropped in a bed of hay. According to predictive processing, my brain ups the precision-weighting on specific aspects of the visual information that would indicate a small silvery object, thereby increasing my chances of success. That's what attention, if these accounts are correct, really is—attention is the brain adjusting its precision-weightings as we go about our daily tasks, using knowledge and sensing to their best effect. By attending correctly, I become better able to spot and respond to whatever matters most for the task I am trying to perform (for more on this, see the Appendix). Precision estimation is thus the heart and soul of flexible, fluid intelligence.

But what happens when precision estimations misfire? This would skew the impact of different bits of sensory evidence, and of different predictions. Precision estimation is the brain's way of telling itself where, and by how much, to place its bets. When this goes wrong, our brains will bet badly: they will misestimate what to take seriously and what not to take seriously, thereby generating false or misleading experiences. This is exactly what seems to be happening with functional disorders. In these cases, unwilled misallocations of precision act as self-fulfilling prophecies. Predictions of pain or impairment become highly overweighted, and those predictions overwhelm the actual sensory evidence, forcing experience to conform to our own hidden but misplaced expectations.

Such effects, we have already seen, are surprisingly common. I experienced just such an effect when I mistakenly seemed to hear the gentle chirping of the bird alarm back in Chapter 1. But when such effects become entrenched, and when they concern important matters such as our own states of pain or disability, the consequences can be devastating. Human experience, in such cases, becomes radically disconnected from the actual streams of bodily and worldly evidence.

There is good evidence that misfiring precision assignments (unusual patterns of attention) play a role in many, perhaps all, functional neurological disorders. For example, simply distracting the sufferer by making them direct their attention elsewhere often makes functional (but not structural) tremors vanish. Patients with these tremors also spend much longer looking at them than those whose tremors have standard causes and they greatly overestimate how often their tremor occurs. When tremors are caused by structural (i.e., standard neurological) disorders, patients' estimates are much closer to the true frequency. This makes sense if it is disordered attention that, in the case of functional disorder, drives the formation of the tremor itself. In these cases, the process of attending to the tremor ups the precision-weighting on the hidden expectation of tremor and this brings the tremor about.

This creates a version of the famous "refrigerator light illusion." You might infer that your fridge light is constantly on just because the light is on every time you look inside. But actually, it is the act of looking (opening the door) that turns on the light. Similarly, you might believe you have a near-constant tremor because the tremor is always there when you pay attention to it. But if the tremor is actually in whole or in part the result of the process of "predicting and attending" itself, that assumption may be wildly wrong.

Hoover's Sign

A classic demonstration of the role of aberrant attention in functional disorders involves the phenomenon known as "Hoover's sign." Named after the American physician Charles Franklin Hoover, this is present when a patient with unexplained weakness in one leg proves able, when their attention is directed elsewhere, to exert normal amounts of pressure with that leg.

The way it works is this. The patient is asked to make a certain movement with their nonafflicted leg while the examining doctor checks for pressure exerted on the other (afflicted) side. In normally functioning individuals (those with neither functional nor structural weakness) there is a kind of crossover effect such that lifting (say) the left leg causes the opposing hip to extend and the heel of the right foot to exert downward pressure. In cases of structural right leg weakness this crossover effect is missing, as the right leg cannot respond. But if the weakness is functional in nature the patient will involuntarily engage the afflicted leg, exerting downward pressure with the heel. The doctor's request diverts attention to the unafflicted limb, thereby revealing the preserved biomechanical ability of the afflicted one.

Explaining this to patients, the neurologist Jon Stone likes to emphasize the difference between their voluntary leg movement, which is severely impaired, and their involuntary (automatic) movements, which are not. What Hoover's sign shows is that the problem is not really with the power of the leg, nor even with the ability of their brain, when distracted, to deploy that power. Rather, it reflects what happens when attention is directed toward using the afflicted leg.

This is a clever test, and it is widely used today (see Fig. 2.2 for an example).

The early (1908) literature on Hoover's sign depicted it as a means of detecting both "malingering" and functional leg weakness. But nowadays, there need be no implication that the patient is "faking it." Rather, what Hoover's sign suggests is that a certain unwilled pattern of expectation and attention—caused, of course, by very real changes somewhere within the brain—may be the hidden cause of the apparent weakness. Another way to think about this is that the absence of disease

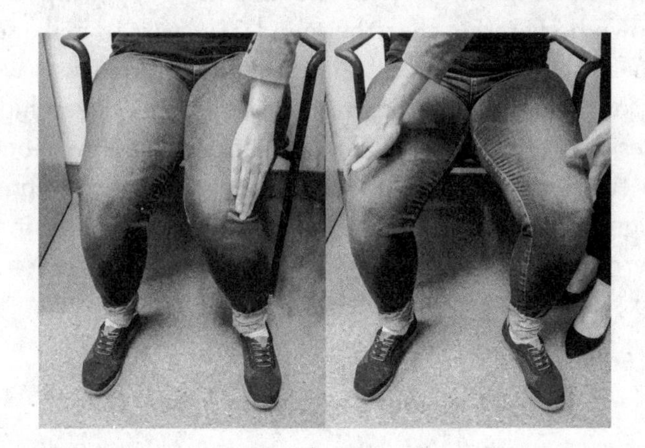

Fig. 2.2 Hoover's sign. Left: Weak left hip extension. Right: Strength in left hip extension returns to normal with right hip flexion.

or gross physical damage does not mean that there is no pathological change at all. But the relevant changes are subtle and deep—they reflect a fault in the complex patterns of signal amplification and dampening that occur within the predictive brain.

Further support for the attentional hypothesis comes from preliminary studies led by my onetime Edinburgh University colleague Professor Rob McIntosh. The studies looked at patients with functional weakness in one (but not both) of the upper limbs. They found that aberrant patterns of attention to the afflicted limb correlated with feelings of numbness. It seems that altered patterns of attention can not only make us experience bodily feelings such as pains when all normal causes are absent—they can also make us experience a lack of sensation (numbness) when no structural cause is present.

Strongly anticipating pain, numbness, weakness, or other symptoms alters patterns of attention (precision-weightings) in ways that can either amplify or entirely generate the experience—which then seems to confirm those very expectations. This is simply our old friend the "White Christmas" effect all over again, but here affecting the way we experience our own body rather than the sounds we hear. To make sense of these self-constructed feelings of pain, numbness, weakness, or paralysis, sufferers may start to suspect deep hidden causes—such as persistent hidden illness. These new beliefs then further reinforce the expectations of those symptoms, reinforcing the cycles of aberrant attention. A similar circular pattern of false confirmation may occur in some cases of psychosis, as we'll later see.

Expectancy and Its Role in Chronic Pain

Functional disorders afford a powerful illustration of the role of hidden predictions and patterns of attention. But there is a deeper insight lurking in the wings. For what all this suggests is that there is no such thing as a raw or "correct" experience of a medical symptom anyway. Since all human experience is constructed from mixtures of expectation, attention, and sensory stimulation, it will never be possible to experience either the world or your own body "as it really is." Indeed, it rapidly becomes unclear what this could even mean. Instead, there exists a deep continuity between cases where expectation and attention create symptoms (as we saw earlier) "from whole cloth" and cases where they also reflect the operation of some more standard form of disease or injury. Functional disorders simply lie at one end of this spectrum.

There is plenty of evidence for this initially surprising claim. Omer Van den Bergh, a health psychologist working at KU Leuven University in the Netherlands, notes that symptoms across a wide range of conditions strongly match their bodily causes only for early, acute, and localized dysfunction or pain—for example, the temporary sharp pains caused by surgery, cuts, and broken bones. Move to chronic conditions and the picture looks very different. For example, reported breathlessness in chronic obstructive pulmonary disease (COPD) shows huge variation for the same level of lung damage, both in different patients and in the same patients at different times. Similar results were found in studies involving reports of atrial fibrillation, asthma symptoms, diabetes, and many more. Multiple studies suggest that asthma patients can often experience symptoms in a way that does not reflect their current pulmonary state but is instead the result of ingrained expectations. Typically, such expectancy-driven attacks—variously estimated to affect somewhere between 15 and 60 percent of sufferers—occur when returning to a context (or encountering a cue) that was associated with a previous attack. This rapidly sets up another self-confirming cycle in which the new attack, being again experienced in that context, seems to confirm the expectation, making such attacks more likely in the future. This is rather like the case of the performer with stage fright whose true abilities are masked by their own mounting expectations of failure. The circularity is daunting. Every new instance of stage fright confirms the expectations of failure, and those expectations ensure that the instances of failure accumulate. Recognizing this circularity is, however, often the first step in breaking the cycle, as we'll see later when looking at ways to "hack" our own predictive brains.

Something similar seems to be occurring in some cases of chronic back pain. In a 2019 interview, the London-based health psychologist Tamar Pincus commented that:

> after several bouts of back pain, people start to process the world differently . . . their pain becomes embedded [among] the things they associate with themselves. If they are shown an image of a staircase, for instance, their first thought is, "I can't climb it." After a while, you see and feel things coated with pain. You no longer need the injury to feel pain. And you might experience more intense pain, purely because you're expecting it.

Individuals will differ in how they assign precision to bodily signals, including those associated with pain and disability. Moreover, living with a condition for a long time enables idiosyncratic expectations (for example, about severity in different contexts) to arise and become ingrained. This means that even where there is some standard structural cause such as a bulging or herniated disc in someone with back pain, the way we experience our symptoms may over time come to involve large doses of mindset and expectation.

In a certain sense, chronic pain at that point is perhaps

best considered not so much as a symptom, but as the disease itself—the very state that needed to be addressed. On the first Global Day Against Pain in 2004, it was declared that "chronic and recurrent pain is a specific health care problem, a disease in its own right." Since then, this once marginal viewpoint has become increasingly influential in both theory and in clinical practice. Predictive processing provides the first fully formulated theoretical framework within which this strong claim can be defended and made precise. It shows us exactly why, as leading pain theorist Mick Thacker puts it, we need to move away from thinking of pain as a simple sensation, a direct signal of damage or potential damage, to a view of pain as a perception. Like all perceptions, it takes shape only thanks to the precision-weighted interaction of predictions and current bodily signals. It is that process of combination that provides the wiggle room that enables persistent pain or impairment without damage, threat, or disease. This is not to say that everything will respond to changes in expectations. It won't. But attention and expectation are key players in the construction of all our experiences of health and illness, and this is true even when standard structural causes (damage or disease) are present.

Altered Balances in Autism Spectrum Condition

Predictive processing also sheds considerable light on a wide range of typical and atypical forms of human experience. A good starting point is to notice that there are two very broad ways for such processing to go wrong. The first is for the brain to underweight predictions and expectations. This will make it hard to detect faint but predictable patterns in a noisy or ambiguous environment. But the second general way to go wrong is for the brain to overweight expectations. In extreme cases, overweighting results in hallucinations. You seem to see and hear things that aren't there, just because—like those phantom phone vibrations or the chirping of the imaginary bird alarm—they are at some level strongly expected.

Autism spectrum condition was initially thought to reflect a specific imbalance of the first kind—a systematic underweighting of prior expectations. This was the "weakened prior" theory of autism. Underweighting prior knowledge would make weak or elusive patterns hard to detect, and hard to learn too. Such patterns would include things like facial expressions, intonation, or body language, things that delicately hint, in context, at other people's mental states and attitudes. An imbalance of that kind would also make it very hard to learn these patterns in the first place, and even harder to recognize them in situations that are complicated or ambiguous. Recent evidence casts subtle doubt, however, on this bald initial hypothesis. Rather than weakened predictions, intriguing evidence is emerging that suggests that the core issue involves (not underweighting knowledge-based predictions but) actively overweighting the incoming sensory evidence.

You might think that these are essentially equivalent—it's a balancing act, after all, and underweighting one side of the scales (predictions) will have many of the same effects as overweighting the other (sensory stimulation). But good evidence against the simple "weakened priors" theory has been found using Mooney images of the kind we met in Chapter 1. Mooney images, you will recall, are simple black-and-white images that are hard to decipher at first, but very easy to see once you have been exposed to the full grayscale image on which they are based. A team of psychologists from the Netherlands showed Mooney images and their source images (the original, non-Moonified pictures) to people with autism spectrum condition. They also showed the images to a large and varied group of more neurotypical participants. All were later shown the Mooney image again and asked to identify what it represented.

Contrary to what would be expected under the weakened priors theory, there was no difference in performance between those with autism spectrum condition and the other participants. The clear conclusion is that the ability to use acquired prior knowledge to perform the Mooney task is intact throughout both groups. This favors an alternative theory in which those with autism are instead overweighting normal sensory evidence rather than underweighting their own knowledge or predictions. Converging evidence now favors the broad idea that a tendency to overweight the sensory evidence (enhanced sensory precision) is the core difference separating autism spectrum condition from the more neurotypical profile. What might this mean in practice?

Enhanced Sensory Worlds

Writing in *Spectrum* (an online forum for autism spectrum condition research news) George Musser reports a conversation with Satsuki Ayaya, a PhD student in Tokyo with a diagnosis of autism. Ayaya reports that her experience presents her with excessive amounts of detail that often do not serve her daily needs. As Musser writes:

> she feels in exquisite detail all the sensations that typical people readily identify as hunger, but she can't piece them together. "It's very hard for me to conclude I'm hungry," she says. "I feel irritated, or I feel sad, or I feel something [is] wrong. This information is separated, not connected." It takes her so long to realize she is hungry that she often feels faint and gets something to eat only after someone suggests it to her.

She doesn't just feel "hunger," instead the more fine-grained specifics of the bodily signals dominate. You are feeling a whole lot of something—but what is it? According to the overweighted sensory information theory, autism spectrum condition individuals constantly encounter an excess of highly detailed and apparently very salient sensory information of this kind, coming from both inside their own body and the outside world. This sensory excess impedes the moment-by-moment identification of the broader context or scenario (in this case, hunger). In other words, the emphasis on every aspect of sensory detail effectively makes it impossible to spot the larger forest for the trees.

Moreover, just as neurotypical people build environments that suit the ways neurotypical brains balance evidence, expectations, and uncertainty, so those with autism spectrum condition structure and seek out environments that better fit their own distinctive inner balances. I was recently invited to an Autism Community Research Network session on the predictive mind. At that session, an individual with autism spectrum condition commented that "if the world was dominated by my cognitive profile, there would only be quiet trains, perhaps with a single designated 'noisy' carriage." In other words, we'd have built a world where the distribution of quiet carriages reflects a different cognitive profile.

Faced instead with an endless stream of rich and apparently attention-demanding sensory information, an individual with autism spectrum condition might start to self-select more predictable environments, becoming increasingly wary of complex social encounters. Repetitive and stereotyped behaviors such as rocking or hand-flapping might also emerge, as these would offer a clever way to ensure (by self-generating) a predictable stream of sensory input. Yet another way to reduce sensory surprises is to develop extreme expertise in a restricted domain. In sum, a lot of otherwise disparate effects fall into place if we think of autism spectrum condition as involving the overweighting of incoming sensory information.

The McGurk Effect

Autism spectrum condition seems to involve giving excess "credit" to the detail and nuance of the stimulations arriving from the senses. But what counts as excessive credit anyway? Giving weight to every nuance in the sensory signal can (in the worlds we mostly live in) be a source of overload. But it can also, at times, reveal more of what is really there. A nice example of this involves an auditory illusion known as the McGurk effect. This effect is best understood as a variant of ventriloquism. In ventriloquism, auditory signals from the ventriloquist are heard as if they were coming from the dummy due to the temporally well-matched movements of the puppet's lips. Once more, our own hidden expectations of appropriate causes mislead us—we subconsciously expect the sounds to be generated where we see the mouth most clearly moving. In effect, the brain infers that the auditory cause is where the puppet's mouth is moving. That strong expectation subtly warps the perceptual experience by causing the brain to underweight real sensory evidence to the contrary (treating it as noise rather than signal). This is the same effect that we already saw in the Hollow Mask illusion (Chapter 1) and other cases.

In the McGurk effect, subjects are shown a video clip where the sound "ba-ba" is played, but the person's lips are actually moving in the ways they would if they were saying "ga-ga." Faced with this apparent contradiction, neurotypical subjects tend to merge the two sources of information, and clearly hear "da-da." The "da-da" perception is a kind of illusion. It is the brain's best guess at what the world might be throwing at it, given the sensory evidence and what it knows about the relation between speech sounds and lip movements. There are many videos online demonstrating the McGurk effect.

The McGurk effect is diminished—and sometimes entirely absent—in those with autism spectrum condition. This makes sense if these individuals take the incoming sounds at something closer to face value ("ba-ba"), rather than warping their experience to conform with the guess that best accommodates the accompanying visual information. Similar reductions in susceptibility have been found with certain other illusions (including the Hollow Mask illusion) too. In such cases, autism spectrum condition results in what is intuitively a more accurate perception, insofar as the perception is less influenced by the illusion.

These questions of what's "truer" to the sensory evidence are, however, remarkably slippery, as we'll see in more detail as our story unfolds. The price of a more accurate perception in one context may be a tendency to make costly mistakes in others. No one way of balancing sensory evidence and prior knowledge is going to be perfect for all purposes.

Altered Balances in Schizophrenia

Predictive processing may also shed some light on a frequently misunderstood condition—schizophrenia. The psychologist Peter Chadwick describes his own experience of the onset of schizophrenia as involving what he called a "step-ladder to the impossible." As he ascended the rungs of the ladder, he ascribed increasing significance to an array of patterns and coincidences, slowly forcing him to shift his fundamental understanding of how the world works. As he puts it, "I had to make sense, any sense, out of all these uncanny coincidences. I did it by radically changing my conception of reality." For example, he started to hear things being said on the radio as if they were spoken directly to him, picking up on what he was already thinking in some inexplicable kind of way. To make sense of what may initially have been a few coincidental and vague such linkages, he then inferred a large hidden conspiracy of technological experts:

> Obviously, there was indeed an Organisation of technological experts, informed by past enemies, the neighbours and maybe by newspaper personnel, out to monitor and predict my thoughts and then send in replies by the radio.

How are we to understand such a process? Schizophrenia often involves both hallucinations (apparent perceptions that fail to match the real world) and delusions—strange beliefs such as the belief in an organization of technological experts. Important early research applying ideas about the predictive brain suggested that these two features might be flowing from a common source: waves of falsely generated prediction error signals.

The idea was that schizophrenia involves the mistaken generation of highly weighted (high-precision) prediction error signals. These would typically signify important yet unpredicted sensory information, such as the multiple "coincidences" of which Chadwick speaks. That information is then propagated deep into the brain, which treats it as signifying important news. This, the authors argued, is akin to that malfunctioning dashboard warning light. It screams "take immediate action" even though there is nothing really in need of attention. Falsely generated, highly weighted prediction error forces the brain to seek a new predictive model. The resulting hypotheses (such as the Organization, telepathy, alien control, and nowadays all manner of strange beliefs involving the internet) may appear bizarre to the external observer.

Yet from within they constitute the best—indeed the only—explanation available.

This also helps explain what may be going on during the distinctive early stages of the onset of psychosis. There, although the person is otherwise functioning much as usual, the world begins to look somehow different or strange. That strangeness, the authors suggested, reflected the presence of persistent, unresolved prediction errors. Those errors then slowly drive the system to form increasingly radical hypotheses in an effort to accommodate them.

Importantly, predictive brains control action as well as perception, and so the delusional person will actively seek out confirming evidence for their radical hypotheses. As this process unfolds, new information may itself be interpreted differently so as to appear to confirm or consolidate the radical beliefs. The cycle of error thus becomes (yet again) viciously self-protecting. Such pernicious outcomes seem to be the Achilles' heel of the predictive brain.

However, it is still early days in this research and caution is warranted. Aberrant prediction errors, even if they play a role in the development of psychosis, are probably seldom (if ever) the whole story. Almost certainly, we here confront a complex mosaic of causes, and there are probably multiple pathways to psychosis each having its own distinctive "feel." Changing cultural context matters too—contemporary delusions, as just mentioned, often involve bizarre beliefs concerning the internet, while more traditional ones (such as beliefs involving telepathy and aliens) seem to be becoming less prevalent. These are all important facts in need of better explanations, and much work remains to be done. But however complex the final story turns out to be, it now seems certain to involve cascading alterations to the delicate system of checks and balances that characterizes the predictive brain.

Post-Traumatic Stress Disorder

Predictive processing is likely to have a similarly transformative effect on the understanding and treatment of post-traumatic stress disorder (PTSD). Sometimes described as a "reality monitoring" deficit, sufferers experience flashbacks that may be hard to distinguish from real events, as well as hyper-anxiety, avoidance behaviors, and a host of life-changing symptoms. It is estimated that up to 30 percent of those who are exposed to a highly traumatic event (such as rape, combat, or domestic violence) will go on to develop some form of PTSD.

In some revealing recent experiments, researchers recruited fifty-four combat-exposed veterans who had a wide range of PTSD symptoms. Twenty-four of these had a diagnosis of PTSD and thirty did not. Participants were exposed to one of two "mildly angry" face images, followed by a mild electric shock. At first, face A predicted the shock one third of the time, while face B never predicted the shock. Then, in a "reversal phase," those associations were reversed, so that the face that was previously most likely to lead to the shock did not do so, while the other one did. The researchers used skin conductance as a reliable physiological measure of when, and how strongly, a shock was expected on each trial by each participant. This is the electrodermal response that occurs when we are stressed or otherwise aroused, because secretions from sweat glands in the skin temporarily render it a better conductor of electricity.

The authors then applied multiple different models to see which one best accounted for the patterns of physiological data as they evolved over these trials. In the winning model, PTSD severity was extremely well correlated with unusually large increases in precision-weighting on the prediction error signal in response to unexpectedly negative outcomes (unexpected mild shocks). In the most severely affected individuals, the response to failing to predict the shock was to radically overweight the missed cue (the specific face) and thereby become hypersensitive to its occurrence in future. In a war zone, such a tendency would result in sustained, perhaps even lifelong, hypersensitivity to the cues (lights, flashes, sudden noises) that presaged unexpected and life-threatening events such as sudden missile attacks, explosions, or other threats.

This may help explain why some people who have suffered trauma go on to develop PTSD while others, in exactly the same situation, do not. Different human brains respond differently to prediction error signals following sensory surprise. As a result, some individuals will be more susceptible to PTSD and other debilitating conditions. If this proves correct, tests like these could one day be used to identify those people who are most at risk, adjusting their roles during military service accordingly. That would be an instance of what has been called "perceptual phenotyping"—the use of psychological tests to help build cognitive profiles that identify individuals at risk. Such tests will also offer important clues to which individuals would benefit most from different interventions or altered environments. I suspect they would rapidly reveal a surprisingly wide range of cognitive profiles within the neurotypical population. Such profiles could one day be used to inform training, learning, and rehabilitation, tailoring each more closely to individual needs and differing cognitive styles.

So Which Balances Are Best?

Predictive brains host complex (precision-weighted) balancing acts, and different balancing acts result in diverse ways of experiencing and responding to our worlds. But what constitutes a good or optimal setting for these various balances—one that will ensure that perceptual experience reveals things "as they really are"? Unfortunately, there is no way even to address that kind of question without making many assumptions about the nature of the actual environment and its relative stability or tendency to change (its volatility). For example, how often in your environment do unexpected loud noises occur without life-threatening explosions, and is that ratio stable or constantly changing? We also need to factor in some kind of risk/reward structure. Just how costly is it, in the environment you happen to live in, to mistake a backfiring car for a sniper shot, or a booming subwoofer for an explosion?

Predictive balances that might save your life in one kind of environment may do you all kinds of harm in another. This means that different ways of balancing predictions and sensory evidence are good or bad only in relation to the kind of world you happen to inhabit. To illustrate, one study put so-called nonclinical voice hearers (people with no diagnosis of psychosis but who often hear nonexistent voices) and people who don't hear such voices into noisy fMRI machines. But they had hidden clips of distorted speech (sine-wave speech, which we met in the previous chapter) in among all that general noise.

In this unusual setting, the voice hearers were at a distinct advantage. Seventy-five percent of them detected and could interpret the hidden speech, compared to only 47 percent of those who didn't tend to hear nonexistent voices. In other words, the voice hearers were a lot better than the rest at detecting the unexpected presence, amongst the din of the fMRI, of real (though artificially degraded) speech. The voice hearers were thus experts at actual voice detection. This makes sense if we suppose that these individuals constantly maintain a high expectation for speech sounds, leading to false positives (illusory percepts, like my phantom phone vibrations) in normal life, but also supporting better detection rates when the sensory evidence is extremely thin.

Now imagine a world in which there are dangerous predators that (rather like comic book villains) tend to announce their arrival with a softly whispered verbal threat. Imagine also that this is a noisy world, full of thunder, bangs, and whistles. In such a world, the prey inhabit a niche where the detection of whispered threatening words against the noisy background is paramount. Hear the whisper and you just might get away. Imagine too that false positives are not costly: mistakenly thinking that you have heard one of these threats won't usually be much of a problem. We have now imagined a world in which the altered balances of the nonclinical voice hearers would be highly adaptive. This suggests—and this is a theme we will return to later—that there simply isn't any single, correct way to balance predictions and sensory evidence. Perhaps the best we can hope for is to be suitably sensitive to changes in the environmental profile as they occur, altering the ways we balance different predictions and sensory evidence accordingly.

•

I've tried to show that we encounter not just the outside world but also our own body and medical symptoms in ways that are strongly shaped by a web of predictions installed by past experience. What we see, hear, and feel—even when everything is working exactly as it should—is never a direct reflection of the state of our own body or the wider world. Instead, the world and body we experience is always part construct: a product of our own conscious and nonconscious predictions. This helps explain many otherwise puzzling phenomena. These include the nature of chronic pain, the origin of functional disorders, psychosis, and perhaps one day the full sweep of both atypical and neurotypical forms of human experience. These phenomena, if our story is correct, all reflect shifting (precision-weighted) balances between predictions and sensory evidence.

The hope is that by better understanding those balances, and the various ways they can go awry, we may begin to move beyond a superficial, symptom-based understanding toward something more principled—a unified approach in which psychiatric and functional conditions are classified according to deep causes. This should one day enable more targeted, individualized, and effective interventions. It should also help us start to see beyond many old and unhelpful distinctions such as the dichotomy between the mental and the physical, and between what is "psychiatric" and what is "neurological" in origin.

But our story is only just beginning. Predictions, as we have seen, shape human experience. But they shape human action too, and it is that crucial role to which we next turn. Actions are a way of making certain predictions come true.

3

ACTION AS SELF-FULFILLING PREDICTION

WE HAVE seen that expectations and predictions deeply influence what we see, hear, and feel. But we have so far said little about how they influence what we do. Yet the fundamental task of the brain is to help us to stay alive, and that means acting in a complex and uncertain world. Action is where the rubber really hits the road—where the high metabolic cost of having a brain must really earn its evolutionary keep.

Perhaps surprisingly, prediction is also the engine of action. This is because ordinary daily actions (according to predictive processing) are caused by predictions of bodily sensation. They are caused, more precisely, by predictions of the flow of bodily sensations that would occur if that very action were to be performed. The predictive control of action thus has a kind of subjunctive quality. The brain predicts how things would look and feel if the action were being successfully performed, and by reducing errors relative to that prediction, the action or movement is brought about. Predicting just how it would look and feel to hit that perfect drive, or make a killer serve, brings that longed-for result about. But this is not a facile nod toward "positive thinking." Rather, it is a detailed proposal about how our brains control our bodies. The upshot is that successful action involves a kind of self-fulfilling prophecy. Predicting the detailed sensory effects of a movement is what causes that very movement to unfold.

By making prediction the common root of both perception and action, predictive processing (active inference) reveals a hidden unity in the workings of the mind. Action and perception form a single whole, jointly orchestrated by the drive to eliminate errors in prediction.

Ideomotor Theory

There is a mode of controlling action that is remarkably well suited to delivering fluid, flexible control. It is also a mode of control that comes very naturally to a predictive brain. Its roots go back to the mid-nineteenth century and the work of the German philosopher Hermann Lotze—work that was then taken up by the American philosopher and psychologist William James. The core idea was that actions come about because we mentally represent the completed effects of the action. In other words, the idea of the completed action is what brings the actual action about. This is sometimes said to reverse a commonsense notion of causality, since instead of the action causing the effect, it is the representation of the effect (the completed action) that causes the action itself to unfold. It's not really that the effect precedes the cause, but rather that the cause turns out to be a kind of mental image of the effect. This became known as the "ideomotor theory of action," since the idea (or mental image) of the completed motor action is what brings the actual movements about.

Let's begin to consider how this might work in practice. Imagine you have a wooden marionette, with multiple articulated joints, that can be animated by a network of strings. You want to make it raise its hand. By pulling the string attached to the hand, you move the hand to where you need it to be.

But in so doing, you automatically move all the interconnected arm, elbow, and shoulder joints in exactly the ways required (see Fig. 3.1). This means that the movement planner (in this case you, the puppet master) need not worry about exactly how to move (say) the shoulder joint or the elbow joint. They take care of themselves, falling into the correct configuration as the hand is drawn toward the desired endpoint.

In slightly more technical language, you don't have to worry about all the many degrees of freedom in the arm, elbow, and shoulder. By solving the problem for the endpoint of the hand, you automatically ensure that everything necessary falls into place. From the point of view of the marionette (assuming it had one) it would feel like its hand was pulled by some external force directly toward some desired location and that the rest of its body simply fell into the shape and form required. This is also known as the passive motion paradigm. This states that the task of the brain, when controlling action, is to determine how each bodily joint would have to move if some external force somehow pulled the body toward the goal. It is then the brain's careful simulation of all the bodily effects expected under that scenario that causes the bodily parts to move in exactly the ways required.

This method of controlling movement involves a profound inversion that will appear again and again as we look at how predictive brains control action. One approach has it that our brains must find the right motor commands by working forward from the actual state of the body toward the target state—computing the complex sequence of commands to take us from "hand at rest" to, say, "hand gripping coffee cup." This is a very complex problem with many possible solutions. Predictive processing suggests something like the reverse. Representing some desired end result, such as grasping the cup, automatically recruits (in the skilled agent) the set of motor commands needed to make that very thing happen.

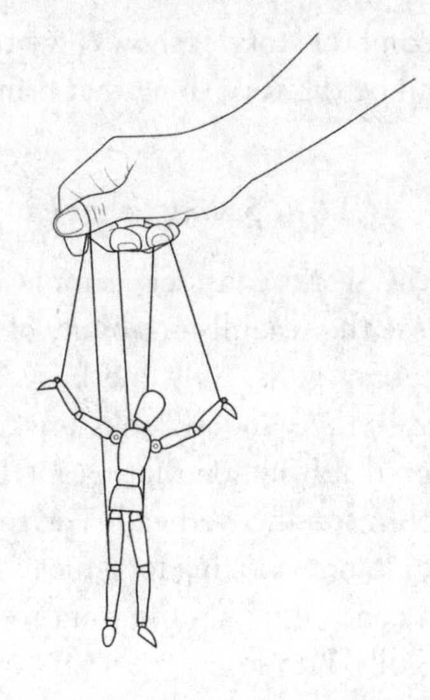

Fig. 3.1 The marionette metaphor of the passive motion paradigm. The "internal model" that coordinates and plans the motion of all the joints operates on a small set of force fields applied to "focal points" of the body model.

How could this possibly work? In the case of the marionette, the puppet master is quite literally pulling the strings. But for this to be a way of understanding how brains control movements, we need to understand how we can be both marionette and puppet master at the same time. Fortunately, predictive brains are ideally suited to enable just this kind of "magic" to occur. In the broadest possible terms, the solution is that the brain learns, through training and experience, to predict what we would see and feel if—but only if—our bodies were moving in just the right ways so as to achieve our goals. Those predictions (of what we would see and feel as the right movements unfold) then act—in a way we are about to explore—as motor commands bringing those very movements about. In this way predictive processing provides a way of carrying out the procedure envisioned, in very general terms, by the older ideomotor story. It shows how the "idea" of a successful action can be the very thing that brings that action about.

Seeing Seagulls

To see how this works consider a simple action such as turning my head to see the seagulls out of my office window. I cannot currently see any gulls, only the busy desktop of my home computer. Still, my window looks toward the swaying masts of boats moored at Brighton Marina and I can hear those pesky gulls squabbling loudly overhead. The sound of the gulls, and the fact that I'm now looking for a nice example of prediction-based action control, makes me want to look out the window and see the gulls. I do so.

In predictive processing terms, what happened is this. The sound of the gulls, and my need for a familiar example, made me strongly predict looking toward the gulls. The best way to get rid of the resulting prediction errors (which were many, since I was still actually looking at my busy computer screen) was to turn my head *just so* and move my eyes *just the right amount*. By predicting the specific sensory effects of that motion, and then rapidly quashing the errors that resulted by actually moving my head and eyes in the right ways, I brought into view the squawking mass of super-sized South Coast gulls.

How did my brain find just the right signals to send to my body to enable all that to unfold? The answer is again by means of a learned predictive model. We can think of that model as the distilled wisdom from prior experience. Courtesy of that distilled wisdom, a predicted effect such as "now seeing the gulls," leads to multiple further predictions of much more specific sensory effects—for example, the ones that would be occurring if my neck muscles were moving in a certain way, the way they would have to move if my head were turning toward the sound. Since my head is not yet turning in those ways, those sensory consequences are not occurring. That delivers a rich stream of prediction errors that are then eliminated by moving my neck muscles so as to make the sensory consequences occur. By launching a cascade of sensory predictions, and then rendering them true by means of action (thus eliminating the resulting prediction errors), the brain creates the desired movements. In other words, I strongly predict looking out the window and that prediction acts as a kind of self-fulfilling prophecy.

In this cameo, sensory consequences are first specified at a very abstract level—something like "I'm turning my head toward the screeching gulls." But as the processing unfolds, those top-level predictions spawned a sequence of lower-level predictions. Importantly, some of these specified what is known as "proprioceptive" sensory information—information conveyed by internal bodily signals (coming from muscles, tendons, joints, and skin) that reflect the position, orientation, or movement of the body in space. At the bottom rung of the ladder lie predictions that engage spinal reflexes that move the body. So the whole process is one in which abstract predictions cause ever-more-concrete predictions, enabling my top-level representation of a desired consequence (seeing the gulls) to cause the bodily actions that bring it about.

The deep unity (under predictive processing) of perception and action should now be apparent. There are two different, but equally effective, ways to minimize prediction errors during our encounters with the world. The first is by using prediction errors to help us discover the best guess about how things are out there in the world. But the second is to act so as to make the world fit some of our predictions. Instead of finding the prediction that best fits the sensory evidence (perception), you now find or create the sensory evidence that best fits the prediction. This is the predictive processing route to action. Actions are simply the brain's way of making its own proprioceptive predictions come true.

One Wiring Diagram to Rule Them All

This also resolves a puzzling physiological anomaly. According to a traditional cognitive scientific account, perception and motor control work in very different ways and move in entirely opposite directions. Perception is the inward flow of sensory information, while action unfolds in the opposite direction. But if perception and action were really constructed by the brain in radically different ways, one might expect a corresponding difference in the directionality and flow of information processing in the brain. Yet surprisingly enough, the wiring diagram of "motor cortex," and the flow of information in that cortex, turns out to be very much like that in sensory (perceptual) regions of the brain. Where we might intuitively have expected motor control to involve a kind of inversion of the wiring diagram for perception, the same flow seems to be at work in both cases.

Predictive processing resolves this anomaly by showing how action and perception can each involve the same kind of wiring and flow of information. Now, motor control works in very much the same way as perception. In each case, the brain is seeking to achieve a fit between what is predicted and what the sensory evidence suggests. But in the case of action, the fit is achieved by altering the sensory evidence to bring it into line with the prediction.

You might wonder how the brain knows which way to proceed on any given occasion. If my hand is (in fact) not yet reaching for the beer glass, why not update the predictions to fit the (correct) sensory evidence that says that my hand is not actually moving? The surprising solution is for the brain—

when I want to move my hand—to gently lie to itself, forcibly downgrading genuine sensory information associated with the currently immobile state of my arm, while upgrading its own prediction of the proprioceptive signature of the grasping motion. In other words, in order to move my body at all, my brain needs to downplay some perfectly accurate information about my own bodily state.

This is achieved by variations in our old friend, precision-weighting—the predictive processing version of "attention." To move my arm, I must give high weighting to the desired future state (arm moving) rather than the undesired present state (arm not moving). That means actively disattending to the present state of the arm, hence dampening down that sensory information. This kind of careful attending and disattending is also familiar from sports coaching, where skilled players are taught to imagine the desired outcome (the spot where the tennis ball will go, the arc of the golf ball to the green).

According to the predictive processing schema, the imagined outcome spawns (in the skilled player) the right set of predicted sensory states, which (since they are not actual) in turn generate a cascade of prediction errors whose fluent, automatic corrections then deliver the stroke, swing, or other action. Just as in the ideomotor story of Lotze and James, it is the idea—the mental representation—of the desired outcome that is crucial to making that outcome real. But for this to work, we need to attend away from the sensory information about our current bodily state and attend toward (increase the precision-weighting upon) the sensory information that would be expected if the action were successfully performed.

This also helps explain why it is that (in the case of already skilled players) simply imagining yourself playing the shots or making the movements can actually improve subsequent performance. Even though no physical actions are involved (this is pure imaginative rehearsal) we are training ourselves to generate the mental representations that, on the day, will actually serve to bring the shots or movements about.

What Tickling (Really) Teaches

This picture of motor action also sheds light on a set of longstanding puzzles concerning most people's surprising inability to tickle themselves. Why is this so hard to do?

Back in 1950 the great German behavioral physiologist Erich von Holst proposed that every motor command given by the brain is accompanied by a second copy of that command, the "efference copy." That second copy, Holst believed, was sent internally to a kind of onboard simulator able to predict (in advance) the sensory consequences of the action. So as soon as we try to tickle ourselves, the simulator circuit has already anticipated the sensory effects of the action. This removes any element of surprise, hence the inefficacy (for most people, under most conditions) of self-tickling. Self-tickling is thus rather like trying to tell yourself a joke—you know exactly what's coming, so the punch line just can't do the usual work. Clever tests of this theory involved the use of a robot tickling device that disrupted subjects' predictions by inserting unexpected time delays. Under these bizarre conditions, we are indeed more easily able to tickle ourselves.

One reason why nervous systems might incorporate simulator circuits is to finesse the problem of time delays. Consider, by way of analogy, the simple task of keeping your balcony garden alive in the hot summer months. One strategy (that I have used too often in the past) is to wait until you see your plants wilting and dying, then rush out with a large container of water. This is not a good strategy. Far better to anticipate the problem in advance, and water the plants every evening. For many processes, waiting for feedback cues (such as wilting plants) is a bad idea, as what is really needed is ongoing preemptive action.

This is also true, as we'll see in the next chapter, of many of our own internal bodily states. We do not wait until we are actually out of fuel (sugar, water) before taking remedial action. Instead, we model ourselves well enough to step in in good time. The same situation arises in nuclear reactors, aviation, and many other areas. In the case of a nuclear reactor, it is crystal-clear that waiting too long for feedback before initiating corrective action is ill-advised. Systems that instead predict the future from their current state, current actions, and a model of how those actions will affect that state, are always one step ahead of the game.

Motor control poses a similar challenge. Due to limitations in the speed of transmission of signals along nerve fibers, real sensory feedback often arrives far too late to be of much use in guiding action. If the brain had to wait for feedback from a moving limb before generating small corrective signals to keep it heading correctly toward some target, the time delays would induce small shakes and oscillations. Indeed, this is exactly what happens when certain parts of the neural apparatus are damaged by stroke or other misfortune.

Thinking about motor control in these ways delivered real insights. But if we understand motor action in the way described earlier in this chapter, this picture gets simpler still. For if brains are prediction machines through and through there is no need for additional circuitry (the "doubling up" implied by efference copy) in order to simulate future sensation. This is because the predictive brain is quite generally in the business of anticipating our own upcoming sensations. That, after all, is how it delivers motor action itself.

In addition, sensory experience should be dampened or attenuated for any body part that is already expected to move even if that movement is being externally generated—

for example by an experimenter who is gently moving your arm while you passively watch. Such dampening, unlike the simple tickling examples, cannot be accounted for by appeal to traditional efference copy (the "second copy" of the motor command). Your brain issued no motor command to move that arm, so there was nothing to copy. But such dampening is indeed empirically observed. The upshot is that there is a general dampening of the sensory impact of expected events quite regardless of how that event (or motion) comes about.

Lessons from the Outfield

By delivering motor control as a result of predicted sensory flows, the predictive processing account also highlights important flaws in the idea (popular in the early days of AI and robotics) of a neat and tidy Sense-Think-Act cycle. Within such a cycle, the main role of sensing was to suck up as much information as possible from the surrounding world, so as to enable the robot to plan actions to achieve its goals. Once the robot had the plan in hand, the role of actions was quite limited—just to carry out the various movements in the projected sequence. This meant that body and action were, in a sense, cognitively unimportant. They were just the way the plan gets executed. But this turned out to be slow, fragile, and unconvincing: a shallow mimicry of the way biological brains control action.

An alternative approach became known as "active sensing." The idea is that sensing itself is an intelligent action, aimed at delivering just the right information, just in time for use. As embodied agents we are able to act on our worlds in ways that actively generate new patterns of sensory stimulation. We move our heads and eyes to explore the visual scene, seeking out subtle cues that will tell us whether that shady form in the alleyway is a dog or a fox. We poke and prod the world around us, to discover object boundaries by seeing what moves independently of what. When a chicken bobs its head, it is actively altering the flow of visual information in a way that helps it determine the relative depths (distances from the eyes) of various objects.

In all these ways, well-timed bodily movements improve our states of information. Perception is no longer a passive phenomenon. Instead, perception and action constantly engage in a kind of coupled unfolding—movements serve up perceptions that enable more motor movements that deliver further perceptions. Vision itself, this body of work suggests, is a highly active and intelligent process.

Bodily movements also transform the landscape for sense-based control. Consider an example from baseball. How does an outfielder catch a fly ball? One thing they don't do is stand there, compute a rich model of the visual scene, work out the entire optimal run trajectory, then deliver the plan fully formed to their waiting body to carry it out. An outfielder, if asked to stand still and just guess where the ball is going to land, will usually do a very bad job indeed. This is because they catch the ball only thanks to a more active strategy that crucially involves their own bodily motions. They run with their eye on the ball, so that their own movement cancels out any apparent changes in the acceleration of that ball as it flies. By running so as to keep the perceived acceleration of the ball in the sky constant, the outfielder reaches the landing spot at the right time to make the catch.

This strategy provably affords a fast, cheap-to-compute way of running to intercept the ball. It is a prime example of embodied problem solving because it makes the outfielder's own movements part of the actual problem-solving process. It is also another example of controlling an action by means of its predicted sensory consequences—the task is solved as long as the outfielder acts to keep their own sensory stimulations within certain bounds. This can be achieved by predicting that the sensory flow will stay within those bounds and minimizing error by moving the body. This is a very robust strategy which automatically compensates for unexpected deviations as might be caused by a sudden gust of wind, since that will immediately cause new and larger prediction errors that will recruit whatever bodily motions are needed to try to counteract it.

Embodied Expertise

To explore the control of action just a little further, think about the way an expert car driver steers through traffic. Perhaps they are in a familiar city with normal traffic but urgently need to get to the airport where their plane is already boarding. They cannot afford to get stuck for long in the lines of traffic and must use all their skills if they are to arrive both safely and on time. A complex sequence of actions unfolds in which the driver seems to simply see where the car needs to go, and at what speed, and performs a linked set of actions that just might save the day.

This requires multiple layers of predictive control. There is one somewhat abstract and distant endpoint in play, which is to arrive at the airport on time. The best way to make that endpoint real is, we infer, to drive unusually fast while still somehow avoiding accidents and getting stopped by the police. That plan (or inferred policy) then gets cashed out, moment by moment, as a flow of local sensory predictions and resulting cycles of correction by means of prediction errors. At every level of processing, the same broad story unfolds. The expert driver's performance is rooted in a constantly changing high-level prediction of how the car needs to behave there and then—one that, because they are an expert, automatically spawns the series of lower-level sensory predictions that act as motor commands, controlling their own bodily responses in the ways needed to bring this about. The net result is to engage a complex set of actions that change gear, steer, and brake, all the while moving their head and eyes in ways that deliver just the right flow of incoming information, at just the right moment to control those actions and responses.

When all goes well, the car then follows the trajectory that the driver was "just seeing." This involves constant subtle corrections that are automatically recruited to eliminate small prediction errors arising whenever the car (as evidenced by the sensory flow) drifts from the correct path. This is the core expertise of the predictive brain. Bodily motions are selected to bring sensory inputs into line with precise top-level predictions about just where the car should now be heading, at what speed, and so on. Those actions, in turn, reflect the overarching policy (drive fast but safely, and don't get pulled over) that looks most likely to get us to the airport on time.

As this process unfolds, the conscious awareness of the skilled driver is freed to be dominated not by thoughts about the unfolding details of their motor action but simply by a kind of "expert seeing"—seeing just where and how the vehicle must move. The prediction-error-correcting brain then does the rest. In this way the car behaves as a kind of extension of the driver's body—a selected trajectory for the car recruits the cascade of predictions of gear-stick motion and foot-pressure sensations needed to bring that trajectory about. In the same way, the brain of the experienced tennis player need only predict the way things would look and feel if the player were engaging the oncoming ball in just the right way to bring about the actions that make that engagement real.

But as all novices know, just wishing that our car or racquet would respond a certain way is not sufficient to make it so. This is because fluent performance of that kind requires a highly trained inner model. Such a model puts perception and action together at every level, from the top (seeing the right trajectory) to the bottom—pressing on the brake just so, while turning the steering wheel exactly the right amount. At every stage, success follows when the inner model delivers predictions that act as motor commands bringing about the predicted sensory states. And as noted earlier, these sensory predictions already factor in all the biomechanics, synergies, and shortcuts that evolution and training have installed. Likewise, when we learn a skill such as driving, what we learn is a structured understanding that is geared to nudging a complex body-vehicle system (with its own complex dynamics) in ways that bring about the predicted sensory states—the ones that would ensue if we were successfully performing the action. In this way, frugality is assured, and perception and action profoundly united.

Such fluidity, efficiency, and success require extensive training. As sports personalities have been known to remark following especially spectacular performances, "the more I practice the luckier I get." The great Muhammad Ali once said, "If my mind can conceive it and my heart can believe it, then I can achieve it." But there is no substitute—as Ali himself knew very well indeed—for hard work and training. The hidden task of all that training, we can now appreciate, is to enable our brains to predict (via a cascade that often starts with a very high-level goal or aim) the many subtle sensory consequences of an unfolding successful action.

This is very different to thinking we need to learn how to "make the action." Instead, we learn how things will look and feel if we are getting the action right—we learn its sensory consequences, highlighting those most necessary for success. Get all that right and predicting those looks and feels automatically controls the body in the ways that most reliably ensure success. The good news is that difficult control problems can, in this way, be solved simply by learning to predict what we would see and feel if we were getting things right. Sensory predictions then act as motor commands, and we find ourselves acting (without knowing quite how we do it) in the very ways that deliver success.

The bad news is that learning just how things would look and feel if we were doing them just right is no easy matter—expert skill still requires long and painful practice!

The Long Game (and the Role of Optimistic Predictions)

We have begun to glimpse what may be one of the deepest implications of predictive processing. It is that goal-directed behavior involves using predicted outcomes to help structure the actions that will best serve to make those outcomes real. In the case of simple bodily movements, such as bringing a coffee cup to my lips, predicting the sensory signature of the desired outcome (the unfolding movement) sets in motion the prediction-error-correcting cycle that brings the desired state (coffee cup at lips) about. This is possible because I know (at some level) how the successful trajectory of action ought to feel. Predicting that feel then acts as a motor program that brings the body into line, making the prediction self-fulfilling.

To guide the coffee cup to my mouth, all I need to do is minimize certain sensory prediction errors in the here and now. But to achieve longer-term goals, I will need to minimize errors relative to the predictions issued by an inner model that looks much further ahead—one that has greater "temporal depth." Equipped with such a model I can act to minimize the errors that my brain calculates would ensue were I not to take a certain course of action in the here and now.

In such cases, the brain is in effect making counterfactual predictions—predictions about what would happen (what I would later experience) if I take, or fail to take, a certain action. This requires knowledge concerning action-outcome patterns that unfold over larger amounts of time. Short-time-scale predictions (such as those that cause my body to move in the here and now) are then nested beneath longer-time-scale predictions about my own future states—such as the state of arriving at that airport on time.

What about even longer-term goals and projects? These are not so very different, except that we must now minimize prediction errors in a yet more abstract and temporally extended space. If I plan to become a better surfer, my brain needs to make the "realistic-yet-optimistic" prediction that I will indeed later be such a surfer. With that goal (long-term prediction) active, I can use what I know about how things work in the world to identify important steps along the way, generating a policy that might—according to my current skill set and personal circumstances—include moving to the coast, taking classes, or vacationing in Tarifa.

Of course, this is not how things seem to me, the person. To me, it just feels as if I want to become a better surfer, and so seek out the actions that will enable me to achieve my goal. But beneath that subjective impression, predictive processing claims, there operates a machinery of potentially self-fulfilling (made true by action) predictions. Predictions like that are somewhat belief-like (being about what is predicted to occur) and somewhat desire-like (being nicely poised to bring those very things about).

These are fully fledged exercises of agency, and they turn on hidden predictions about what I will or will not experience in the future if I do or fail to do such and such in the present. That requires a temporally deep prediction-issuing model, and one that includes me as an agent within its scope. That model delivers predictions about the varying shape of my own possible futures given different sequences of actions. At the bottom of all this lie predictions about what I will sensorily experience (e.g., arriving or not arriving at the airport on time) if I perform certain sets of actions.

The common idea, taking us all the way from short-term motor control to long-term goal-directed action, is that we are pulled along by our own highly predicted future states—such as the state of drinking the coffee, arriving at that airport on time, or improving my surfing skills. This in turn requires a kind of informed optimism. We must at some level strongly predict that we will occupy the states that we can plausibly attain and that best realize our goals. We will then act in ways designed to eliminate errors calculated relative to the optimistic-yet-realistic prediction that those goals are achieved. Realistic optimism is thus the order of the day.

•

Predictive processing offers an elegant and cohesive picture of perception and the control of action—one that can ultimately take us all the way from seeing a coffee cup, to reaching out and grabbing that coffee cup, to pursuing the complex sequence of actions needed to bring some major life project to completion. By making everything revolve around interacting sets of (precision-weighted) predictions, we reveal an unexpected unity binding perception, action, and long-term goal-directed behavior.

Predictive control of this kind is where perception, action, and worldly opportunity meet. Mind, body, and world have never been closer. But they are about to get closer still.

4

PREDICTING THE BODY

TAKEN AT face value, good prediction seems an unlikely goal for living beings. If the goal is simply to drive down errors in prediction, why don't we settle for a very boring and uninteresting life indeed? Worse still, why not choose something 100 percent predictable but quite rapidly fatal—why not seek out a dark empty corner and just stay there until you die? This is the so-called Dark Room puzzle.

Superficially at least, the contrast seems stark. Human life (and much of the life of nonhuman animals too) displays a general striving toward novelty, pleasure, exploration, and fulfillment. My cats seem to enjoy some kinds of surprising and unpredicted events such as the unpacking of a new kind of catnip mouse. I have been known to seek out the thrills of a fairground roller-coaster ride or the challenge of a piece of experimental theater. There doesn't seem to be anything odd or self-contradictory about the notion of a "nice surprise." But if prediction error minimization is always the goal, how do we explain these behaviors?

Our striving to discover, to play, and to explore is laden with emotion and feeling. There is indeed something pleasurable about the right kinds of surprise. The simplest way to accommodate all this would be to add something entirely different to the mix—to argue that sentient agents are driven by deep motivational forces over and above those of successful prediction.

To take that route would be to miss out on a golden opportunity. It is an opportunity to explore an even more profoundly unified picture, one that reaches out to every aspect of a human life. The key to this will be to see that some important predictions concern the inner states of our own body, such as our heart rate, and the state of our internal organs. By turning the prediction machine inward, we start to glimpse the shape of a truly unified science of mind—one in which emotion, motivation, and the attractions of novelty and exploration fall neatly into place.

At the heart of that unified picture lies a simple but transformative fact. It is that the primary task of all the prediction machinery in our heads is to help us stay alive. A major part of that staying alive involves keeping our own inner bodily states within the surprisingly tight bounds of biological viability. That means acting and responding in ways that help create and maintain the many inner physiological states essential to our continued existence as a living organism. To do this, our predictive brains must also target and proactively control a variety of crucial internal physiological conditions.

To understand emotion and valence, we will need to explore the predictive brain as it seeks to anticipate not just an external world but also an internal world involving the core bodily states necessary to our survival and flourishing. It is embodied prediction of this inward-looking kind that enables us—or so I'll argue—to *care* about our worlds and choices. It is also the reason we need not fear the allure of that dark and deadly room.

Escaping the Dark Room

The most dramatic version of the Dark Room scenario might be called the Simple Death Trap. Why don't we simply find a dark corner (providing fully predictable, meager, unvarying patterns of sensory stimulation) and stay there, slowly growing weaker and then dying?

The Simple Death Trap version has a standard, but somewhat underspecified, solution. It says that creatures like us simply do not *expect* to sit still and starve in dark corners. Even well-adapted darkness dwellers (troglodytes) would predict motion, foraging, and feeding, and those predictions would, as we saw in the previous chapter, become self-fulfilling by recruiting the actions needed to make those predictions—of feeding and foraging—come true.

Someone might reasonably worry that the notions of prediction and expectation are here being stretched beyond their proper limits. Do we really predict constant supplies of food and water, and are we really surprised when, halfway through a long local famine, such supplies are no longer to be found? Just saying that we humans congenitally "predict" exploring and finding food can seem like a shallow, crude, and ad hoc way of responding to the challenge of the Darkened Room. Dig a little deeper, however, and we can discover some important ideas capable of fleshing out that broad brush-strokes response. These ideas go back to the heyday of cybernetics, and concern the twin pillars of living organizational forms known as homeostasis and allostasis.

Homeostasis (deriving from Greek words meaning "same" and "steady") implies a tendency to return to a state, or to return to a place within a range. It is to maintain a viable set of internal conditions despite external fluctuations. This general idea has been traced to the nineteenth-century French physiologist Claude Bernard, though the term "homeostasis" was first introduced in the influential works of an American physiologist, Walter Bradford Cannon, in the 1920s and 1930s. However, it was only with the "cybernetic revolution" of the 1940s and 1950s that the idea really started to gain traction.

Desert lizards provide a simple example. Cold-blooded animals such as these can't regulate their own temperature internally and so must constantly move around so as to regulate their body temperatures, seeking out sun or shade accordingly. Other animals (ourselves included) rely heavily—though by no means exclusively—on multiple systems of internal thermoregulation. In humans, complex control systems involving dedicated brain regions (such as the hypothalamus), a wide range of temperature-sensitive nerve cells, and copious sweat glands all contribute to this process. Thermoregulation in humans is just one aspect of the rich web of inward-looking control and regulation known as "bodily homeostasis."

In the early days of cybernetics, self-regulating systems of these kinds were mostly associated with the use of negative feedback—feedback that acts to return a system toward some target stable state. Negative feedback, however, is not the whole story. There soon emerged a slightly more general concept, allostasis, which is arguably even more fundamental. Where homeostasis implies constantly returning to some fixed point, allostasis highlights the need to alter the fixed points themselves so as to adapt to changing needs and environments.

This is exactly what we humans do when we respond to worries and threat by increasing cortisol levels. This so-called stress hormone floods the bloodstream with glucose, powering up our cells so we can take effective high-cost actions such as running fast. But calling it a stress hormone gives cortisol a bad name. It is really an essential part of the system that prepares us for the kinds of actions our brain predicts are on the near horizon. Rather than waiting for something to go awry, then using negative feedback to bring things back into line again, predictive models allow us to make early, preemptive responses. This is why we feel the need to eat and to sleep long before our blood sugar levels drop too low, or our energy resources become really run-down. Feelings of hunger and tiredness are for the most part signals not of present needs but of predicted, impending challenges: challenges that left unchecked would lead us too far from the safe bounds of physiological viability and health.

To make homeostasis and allostasis possible, networks of inward-facing sensory apparatus inform the brain about the states of guts, heart, viscera, muscular and air-supply systems, blood sugar level, body temperature, and a great deal more. These states need to be proactively maintained just to keep the living being in existence. This is the "interoceptive" (inward-looking) sensory system and it is distinct from, but interacts with, the "exteroceptive" (outward-looking) sensory system that includes vision, touch, and hearing. Both these systems are also distinct from the proprioceptive system—the one implicated (Chapter 3) in motor control—that informs the brain about the position and orientation of our bodily parts in space.

Information about our internal physiological states plays a large role in determining how we should act. When we look likely to stray from the normal conditions of staying alive (organismic viability) interoceptive prediction errors result, and these act so as to bring about actions and responses—such as sweating, or seeking out food—that should help avert the danger. The athlete perspires, and the academic leaves the life-threatening Dark Room and heads for the restaurant.

Curiosity and Prediction Error

But we humans do not merely avoid those deadly darkened corners. We also feel positively driven toward life-enhancing activities such as play and exploration. As we all know, feeling good tends to favor exploration and engagement, whereas feeling bad tends to favor retreat and conservation.

One of the many ways in which nature has contrived to push us toward just the right amounts of openness and exploration involves another important dimension of the internal predictive economy—the brain's estimation of its own "error dynamics." This can sound a little daunting, but the idea itself is relatively straightforward. Estimations of error dynamics track how well we are currently doing at minimizing prediction error versus how well we (our brains) were expecting us to do. In other words, are things going better or worse than expected? Things are going better than expected if we are fluently eliminating more (and more important) errors than anticipated. Things are going worse than expected if we are encountering more errors, or dealing with them less fluently, than anticipated. The feeling of "being in the zone" in sports reflects unexpectedly good error dynamics of this kind.

Importantly, creatures sensitive to their own error dynamics will automatically seek out good learning environments, preferring ones that are neither too predictable nor too unpredictable. In the former case, there's nothing much to learn, so the error dynamics are flat. In the latter, learning is too hard, and errors are not eliminated. Good error dynamics arise in between these extremes. Positive and negative moods and feelings are thought to be reporting these important error dynamics, bringing them forcefully to our attention by making some events and situations simply strike us as way more pleasant and fulfilling than others—a good day on the tennis court, or in the office, when you are really "in the zone" for whatever you are trying to achieve, versus a day when every minor task seems like an uphill struggle.

It is no great surprise that evolved creatures prefer environments that enable good error dynamics. These are, after all, conditions in which we are (quite literally) exceeding our own expectations. Babies and toddlers offer a good example. The places even seven- to eight-month-old infants look, and the time they spend looking at them, represent a kind of "Goldilocks zone"—a sweet spot where they can explore events presenting an intermediate degree of predictability. Within the sweet spot, events are neither too easily predictable, nor too hard to predict. They may look long and hard at a revolving mobile sculpture while ignoring much other surrounding complexity. The upshot of this tendency is that babies spend their time on tasks that deliver learning at a good or better-than-expected rate. This means confronting—and sometimes actively creating—a kind of controlled uncertainty. A toddler may try to build a Lego tower that is just a bit bigger than the last one they managed. Babies, infants, and toddlers all seek out and prefer "just-novel-enough" situations—the ones on the edge of their competencies and understandings that will deliver the right kind of learning opportunity.

These tendencies, and their underlying mechanisms, have also been studied using computer simulation in which some (but not all) simulated organisms were programmed to prefer situations in which greater than expected amounts of prediction error are being resolved. These studies (conducted under the appealing banner of work on "artificial curiosity") showed that the simulated animals that were drawn to the Goldilocks zone consistently outperformed rivals lacking that inbuilt drive. This was especially true in simulated environments where change and volatility were the norm. This makes good sense since dealing with such environments puts a premium on rapid learning and cognitive flexibility.

Being natively attracted to environments in which greater than expected amounts of prediction error are resolved is a neat way of ensuring adaptively beneficial tendencies toward play, learning, and exploration. Such creatures cannot help but seek out and prefer those parts of their world in which useful learning is currently possible. They are not at all attracted to those darkened rooms with their fully—and boringly—predictable profiles. Instead, they will constantly seek out richer environments on the edge of their current knowledge and abilities.

Predictive Body Budgeting

The allure of good error dynamics goes some way toward explaining how it is that predictive brains create patterns of positive and negative affect, actively drawing us toward some kinds of situations and environments while rendering others aversive. But there is more to emotions and feelings than error dynamics alone. Another crucial ingredient involves the various ways that information and predictions about the external world interact with inward-facing bodily information and predictions. In this meshwork of inner and outer prediction lie new clues to the nature and origins of emotion and feeling.

In her groundbreaking book *How Emotions Are Made*, the psychologist and neuroscientist Lisa Feldman Barrett captures the bedrock role of predictions in maintaining a viable bodily state using the compelling metaphor of "body budgeting." Just as a financial budget tracks income and expenditure, a body budget tracks and anticipates the use and replenishment of key resources for maintaining bodily life and functioning. These resources include water, salt, and glucose. To renew them, we engage in familiar activities such as finding and consuming food and sleeping. Allostatic mechanisms are vital to this process.

If we feel thirsty, Barrett notes, we may take a drink of water. We immediately feel less thirsty, even though it will actually take the water around twenty minutes to reach the bloodstream and deliver the required effects. Yet the brain delivers the sensation of a "quenched thirst" right away. You (your body) can afford the wait since the sensation of thirst was activated in advance too. In other words, both the feeling of thirst and the feeling of having quenched your thirst each reflect anticipatory processing. My two cats, Borat and Bruno, are also busy body budgeters. When they detect the telltale signs of an imminent visitor (we suddenly tidy up, put wine in the fridge, and generally fuss around) they immediately seem to feel extra hungry, and demand more food. Their brains have learned that we tend to slip up regarding their usual feeding times when we have guests, and (I imagine) they now start to feel extra hungry in advance, proactively body budgeting for the future.

According to Barrett:

> Every thought, memory, emotion, or perception that you construct in your life includes something about the state of your body. Your interoceptive network, which regulates your body budget, is launching these cascades. Every prediction you make, and every categorization your brain completes, is always in relation to the activity of your heart and lungs, your metabolism, your immune function, and the other systems that contribute to your body budget.

The whole of our mental lives, Barrett argues, reflects nothing so much as the brain's busy and deeply anticipatory body-budgeting activity. To enable these kinds of anticipatory control, the predictive models sculpting human and animal behavior need to be as much inward-looking as outward-looking. Every brain region that has been implicated as central to the generation and experience of emotion turns out, Barrett powerfully observes, to be a body-budgeting region.

This will be our route to better understanding emotion and its links to the predictive brain.

Embodying Emotion

It has long been speculated that bodily signals (reflecting things like heart rate, blood pressure, and arousal) must play some key role in the construction of felt emotion. The great American philosopher and psychologist William James in his 1890 opus *The Principles of Psychology* famously argued that our feelings and emotions are in fact nothing other than perceptions of our own varying physiological responses. According to James it is our perception of the bodily changes characteristic of fear (sweating, trembling, etc.) that constitutes the very feeling of fear, giving it its distinctive psychological flavor.

A popular (and still useful) way to think about James's proposal is to see it as suggesting a kind of "subtraction test." This is a thought experiment in which you are invited to try to subtract all the bodily stuff (your own racing heart, etc.) away from the emotional experience, and ask yourself what would be left? James's claim is that you would be left with nothing that is worth counting as an experience or emotion. What an emotion really is, James's argument suggests, is the self-perception of changes in our own bodily states. But James's story doesn't quite work out. If fear is constituted in the body as a racing heart and trembling hands, how is it different from anxiety? James's account leaves us looking for a simple mapping from each distinct emotional state to a matching, equally distinct, signature in the multidimensional space of inner physiological signals. However, no such simple analogs exist.

On the contrary, large and convincing studies that statistically combine the results reported by multiple experiments find no neat, recurrent "bodily fingerprints" for the different emotional states we seem to experience. There is no single set of bodily responses that is unique to sadness, or shame, or any of the many emotional states we name in daily life. Instead, emotional experience seems to be constructed, moment by moment, from a mixture of cultural influences, evidence and expectations about my current situation and my own current bodily states, and my own idiosyncratic tendencies ("individual differences," to use the catch-all term from psychology). It is this melting pot of influences that the predictive engine inside our heads is seeking to master, when it delivers an experience that I might label as "feeling sad" or "feeling anxious."

Brains master the melting pot by commanding and combining predictive knowledge concerning the inner states of our own bodies, our current and upcoming actions, and the wider world. This takes us way beyond the old idea of simple physiological signatures for different emotions and into the exciting research arena dubbed "interoceptive predictive processing." The central idea is that a single kind of process combines inner and outer sources of information, generating a context-reflecting amalgam that is experienced as emotion. For example, a fast-beating heart will have a very different emotional impact on a person who ascribes the cause as recent exercise versus one who fears they are having a sudden heart attack. The very same bodily information can thus feel very different according to how we represent the larger context in which the bodily signals arise.

According to interoceptive predictive processing, feelings and emotions are what result when we integrate basic information about bodily states and general arousal with higher-level predictions of their most probable causes—for example, heart attack versus exercise. This is simply an inward-looking bodily version of the kind of effect we met already in the early chapters—for example, hearing "White Christmas," hearing the words in sine-wave speech, or decoding Mooney images. What we see, what we hear, and the way we currently experience our own bodily states are all complex constructs—mental phantoms shaped and formed by a mixture of sensory evidence and our brains' best attempts to predict that evidence using everything it knows about the wider world.

Inside the brain, the anterior insular cortex (AIC) is remarkably well positioned to play a major role in mediating such a process. This part of the brain is at the core of a dense web of connectivity that allows it to integrate multiple sources and types of interoceptive information. It has been centrally implicated in the construction of emotional awareness of many forms, ranging from basic emotions to (more on this in Chapter 7) sudden insight, hallucinogenic experiences, and feeling at one with the universe. According to interoceptive predictive processing, emotions and feelings reflect a process that combines interoceptive (inward-looking), proprioceptive (action-guiding), and exteroceptive (outward-looking) information with model-based predictions of all those signals as they are occurring.

The winning predictions will be the ones that best "make sense" of that large and varied body of information. In the "racing heart" case just imagined, interoceptive information about heart rate and shortness of breath is integrated with information about the larger context (working out in the gym), delivering new predictions that may cause us to take a break or grab an energy drink. But alter the context and the very same raw bodily information might cause us to suspect something far more sinister and to dial 911. In other words, it is the predictions that best accommodate both what we know about the larger context and the current mosaic of sensory signals that determine how we feel and how we act. What results is an overall sense of how things are in body and world.

We can contrast this picture not only with the simplistic one-to-one accounts mentioned earlier, but also with rather more sophisticated accounts involving a distinct stage of "cognitive appraisal." These quite popular accounts depict experienced bodily sensations as later combining with appraisals of their significance. Such accounts suggest a kind of two-stage process. They depict the bodily feeling as the evidence, and the emotion as reflecting a kind of higher-level judgment about what it means.

The difference between these views and predictive processing is subtle, but important. Predictive processing suggests a much more thoroughly entwined process in which the way your body feels to you is itself altered by what you know about the overall context. This is because all those sources of information and evidence (raw bodily signals plus all the knowledge you are bringing to bear on the situation) mesh together, feeding influence back downward and impacting neuronal processing at all stages. In this way, even your bedrock bodily sensations may be altered by the way they are currently being framed by your own higher-level thoughts and ideas.

The power of framing was already seen in the case of pain (Chapter 2) such as that experienced by the construction worker who falsely believed a nail had penetrated his foot. We shall see more positive versions of such effects in Chapter 7. But the notion of framing, though useful, can also be subtly misleading. For the framing now actively alters the feeling itself, it does not simply put it in context. On this account we are never simply interpreting some kind of "raw feeling" or emotion. Instead, what so often seem to us to be raw feelings or emotions are in fact already highly informed guesses about how things are: guesses that are based (even though we are seldom aware of this fact) on a surprisingly wide range of evidence, expectation, and information.

Wiring the Mesh

This deep meshing (of multiple kinds of information and influence) reflects a special kind of neuronal organization—one that departs quite radically from a once dominant picture of the evolution of the brain. According to that once dominant picture, the human brain evolved in a largely linear and steadily incremental manner, with more recent cortical and neocortical areas progressively overlaying and controlling older more primitive ones. Human rationality, it was supposed, emerged as the evolutionarily more recent neocortex exerted increasing control over swaths of ancient emotional-instinctual circuitry. In just this way, the great Russian physiologist Ivan Pavlov (owner of the famous salivating dog) thought that the cortex was mostly in the business of inhibiting the primitive emotional responses that would otherwise be launched, reflex-like, by more ancient subcortical mechanisms.

This view of the brain produced a long-standing tradition in cognitive neuroscience of characterizing "higher" cortical circuits as controlling and inhibiting the "lower" subcortical circuits. But contrary to that neat, incremental view, the cortex is not a newcomer to human brain evolution. It has in fact long been part of the basic mammalian neural floor plan. Moreover, both cortex and subcortex have continued to change throughout human evolution. There is growing evidence that cortical and subcortical areas evolved in a highly coordinated fashion that produced rich interdependencies. What resulted is a complex looping arrangement linking cortical and subcortical structures in a web of continuous two-way influence. In these tangled webs, each element is constantly affecting, and being affected by, the others.

It is this looping circuitry that keeps our higher-level prediction machinery in direct contact with our own unfolding bodily states, actions, and their worldly consequences. To take a single example, consider the basal ganglia. The basal ganglia are an ancient group of structures involved in basic motor function. But they are connected to the cortex by at least five separate recurrent circuits. These allow information flowing from cortical areas downward to basal ganglia to return back to the same cortical area. The moment-by-moment control of action relies upon their tight coordination. These cortico-subcortical loops also play a role in the ongoing assignments of precision-weighting, constantly conveying updated information about the state of the body, its readiness for action, and the changing reliability of the bodily information itself. Because so many subcortical circuits are tightly coordinated with internal bodily processes (vascular, visceral, endocrine, autonomic), all manner of information from the body becomes positioned to play a much more important and ongoing role than was assumed by the older "corticocentric" vision of the brain. Thanks to these constant looping dynamics, body, brain, and world become equal partners in the construction of thought, experience, and action.

Seeing from the Heart

Next, consider a long line of experiments starting in the 1960s. In the experiments, arousing images were shown to college students who were then asked to rate the attractiveness of the person shown. The students (who were male, heterosexual) rated images of naked women, while experiencing auditory feedback that they were told—though this was not always true—was reflecting their own heart rate: so a faster beat meant, or so they thought, that their own heart was beating faster. Intriguingly, whenever the experimenters induced a mismatch between the actual heart rate and the auditory feedback, the images being viewed were rated as more attractive.

This may seem puzzling at first, but the finding starts to make sense if we consider that the brain should already be quite good at predicting the actual heart rate, so the false feedback leads to prediction error. That prediction error causes the subject to attend more strongly to the stimulus, making them experience it as somehow "important." It is that added salience that is then reflected in the inflated attractiveness ratings.

In more recent studies, participants rated images of faces as angry, neutral, or sad. They were again provided with auditory cardiac feedback. Sometimes this feedback was correct, tracking the actual heart rate, and at other times it was misleading (false). The experimenters found that when the false feedback suggested an increased heart rate, neutral faces were experienced as looking more emotionally intense. In another important series of experiments, an emotion-inducing stimulus (e.g., an angry or scowling face) was presented in a way that kept the image from reaching conscious awareness. At the same time, a visually neutral face was shown to the subject. Under those conditions the visually neutral face was seen as having a less trustworthy look, looking more "as if that person might commit a crime." By contrast, when paired with a happy face (again flashed below conscious awareness) the visually neutral face was no longer deemed suspicious or threatening.

Importantly, the below-conscious-threshold information (the angry face) altered bodily activity, increasing heart rates and galvanic skin responses—the electrical conductivity of the skin, which tends to increase with sweating and affords another physiological sign of arousal. The predictive brain then treats these physiological signals as further evidence upon which to base the predictions that deliver the conscious perceptual experience, adjusting its overall "best guess" accordingly. Taking all things into account, the best overall guess becomes something like "there is a face out there, and a subtly threatening one at that." This again suggests that how we quite literally see the world and other people reflects a deep and continuous combination of inward-looking bodily and outward-facing worldly information. The resulting perceptual experience reflects the visual features of the face, but in a way that is influenced by information about heart rate and other bodily signals.

In Chapter 5, we'll touch on some of the social and political arenas in which these kinds of body-based effects really matter. Such effects also seem to underlie certain patterns of delusion, adding a further layer to the picture sketched in Chapter 2. Thus consider the bizarre case of Capgras delusion. This is the belief that your loved one has been replaced by an impostor. The delusion seems to be triggered when—for whatever reason—your own body ceases to respond in the usual way to the loved one's presence. These missing bodily responses, such as a slightly raised heartbeat or increased galvanic skin response when in the loved one's presence, are not consciously registered. But their sudden absence again acts as evidence that the ever-whirring predictive brain needs to explain. Meshing in the new evidence, the Capgras sufferer's visual and auditory experiences become subtly reconfigured. Perhaps the person's smile now seems slightly different, or their voice sounds a little higher. Such effects flow, as the authors of a recent study suggest, from the lack of the predicted physiological responses to the person's presence. But the subtly altered visual and auditory experiences that follow then put the Capgras patient in the strange position of seeming to have gathered additional perceptual evidence that something important has changed. The loved one "feels different" and they also look and sound subtly different. This plausibly sets the scene for the emergence of the full delusional belief that the loved one has been replaced by a similar-looking (but not quite perfect) impostor.

Depression, Anxiety, and Bodily Prediction

For a more familiar example, consider states such as depression and anxiety. An intriguing suggestion, again from Professor Lisa Feldman Barrett and colleagues, is that depression is often best seen as a "disorder of allostasis." This would mean that depression involves mistaken forms of bodily prediction involving energy regulation.

Thus, suppose that your bodily internal self-monitoring and energy-budgeting system is somehow malfunctioning. Under such conditions you will under- or overestimate your body's present and future needs. You would then be budgeting badly, storing up or using energy in highly inefficient ways. Sudden waves of unexpected tiredness might then be punctuated by equally unexpected short-lived bursts of enthusiasm and energy. Our bodies' energy budgeting can also be impacted by air travel, by exercise, and by bereavement and loss. There is, at the very least, a complex two-way street—often mediated by predictions of energetic need—linking the mental and the more standardly physiological. But really, this is just more evidence that (as we started to see back in Chapter 2) the old distinction between the "psychiatric" and the bodily/physiological/neurological needs to be abandoned. This unifying perspective also makes sense of the finding that chronic depression involves abnormalities not simply of "mood" but also of sleeping-waking cycles, and of metabolic and immunological response. Tying all these together, Barrett suggests, may be a "central problem with inefficient energy regulation."

Mistakes in energy budgeting would normally be corrected by prediction error signaling—if your brain expects the body to require extra energy in the near future and the expectations prove wrong, prediction errors would normally arise and signal the difference, allowing the brain to update the long-term model that made the erroneous predictions. But among the most notable and devastating characteristics of chronic depression, anxiety, and many other psychiatric conditions is their surprising resistance to new information. This suggests that where such conditions really take hold, there is also a problem with either generating or learning from the prediction error signal.

This inability to learn from prediction error results in the situation that Barrett and colleagues describe as that of a "locked-in" brain. From the perspective of predictive processing, the "double whammy" of poor learning and poor energy regulation makes sense if we suppose that the core underlying issue is aberrant precision-weighting. Overweighted expectations and underweighted new information would result in a kind of permanent or semipermanent lock-in of the existing model, leading us to continue with depressive behaviors that actually serve to reinforce the bad model, and that lend false justification to our prior expectations. For example, we expect not to go out and explore new opportunities, leading us to stay home, and then find that new opportunities (as predicted) keep on failing to present themselves. Hidden within such a familiar cycle may be various failures of bodily prediction involving imprecise interoceptive signals making it hard to correctly estimate bodily needs and hard to update those estimates as prediction errors begin to emerge. Negative affect and fatigue would follow as the body responds by producing "sickness behaviors" designed to conserve energy.

These are just broad brushstrokes of some of the existing proposals linking depression and anxiety to disturbances to our bodily predictions. But there will also be—entirely consistent with this—strong psychosocial processes at work in many cases of depression and anxiety. For example, suppose that you experience a succession of unexpectedly negative social events—your partner leaves you, you argue with your boss, a neighbor complains to you. These all result in "social prediction errors" (errors in the prediction of socially important events). You may start to compensate by mistakenly increasing the weighting on small social cues, including all the many signals (facial, verbal, and those involving body language) that help us navigate stressful or important social situations. Faced with all that extra noise, now masquerading as information, you may start to adopt what has been described as a kind of "better safe than sorry" strategy so as to avoid most social interactions, since their outcomes seem increasingly unpredictable.

This is (of course) a foolproof, though ultimately highly counterproductive, way to reduce uncertainty and prediction error. It is in many ways the closest real-world analog to the classic Dark Room scenario. If you seldom place yourself in challenging situations, you will certainly reduce or eliminate many sources of unexpected prediction error. New higher-level explanations (your neighbors all secretly hate you, the boss probably didn't want to hire you in the first place) may then kick in. The result is a familiar pattern of anxiety-inflated responses to small perceived slights, and then protective disengagement combined with new and increasingly negative images of ourselves and our relations to others. Such maladaptive patterns are often seeded by early experiences such as abuse or neglect that lead us to believe that social rewards are unlikely and social outcomes unpredictable.

Immunizing Ourselves to Positive Information

One important and consistent finding in this area is that chronic depression involves a resistance to updating our negative expectations when confronted with what ought to be good evidence of positive outcomes. This failure to update in the face of good evidence (Barrett's "locked-in brain") most likely involves abnormally high precision on prior negative beliefs, which in turn robs unanticipated positive information of the power to alter the inner model that is delivering negative anticipations. The highly weighted (hidden) belief that outcomes will be negative acts as what has usefully been described as a kind of "cognitive immunization" to the effects of countervailing positive information, causing us to either avoid gathering, ignore, or otherwise downgrade perfectly good positive evidence—such as genuine evidence that we are liked and valued. The immunization scenario appears frequently in contexts where psychotherapists put patients in situations designed to disconfirm negative expectations, only to find their efforts immediately invalidated by the patient. Typical invalidation strategies might include declaring these new positive experiences as exceptions to the rule or insisting that "you, as a psychotherapist, are only friendly with me because you are getting paid for it." They may also take more general forms such as saying "although I succeeded in this exam, in other, much more important exams, I will fail."

Confirming this, in multiple experiments involving both healthy controls and depressed patients, the depressed patients showed very different responses to new positive information. Where healthy controls rapidly updated their expectations in the light of new positive evidence, depressed subjects persisted with their original low expectations. In one set of experiments subjects were told they would be given a difficult test (in fact, it was one cleverly designed to have few clearly correct or incorrect answers). Afterward, they were told they had done badly, as expected. After forming negative expectations for performance on further such tests, the feedback was switched to indicate that they had done better than expected. Healthy controls, but not the depressed participants, rapidly altered their own expectations for future performance.

A second experiment then targeted the immunization strategies that might be at work in the depressed cohort. They did this by telling subjects that the test they were taking when the unexpectedly good outcomes ensued was a well-established, solid, highly reliable indicator of ability in this area. This worked against their deep-set tendency to find alternative and more negative explanations for an unexpectedly positive outcome. In this condition they updated their expectations more like the healthy controls. This begins to suggest (and see Chapter 7) practical ways forward. Therapists and clinicians should explicitly target not just negative expectations but also the complex patterns of downgrading or rejecting new positive evidence that result from the abnormally high weighting placed on prior negative beliefs.

These remain very general descriptions, consistent with many theories about the nature and roots of anxiety and depression. The distinctive contribution of the predictive processing account is that it is capable of explaining these general behavioral patterns by positing specific forms of disturbance to the brain's underlying computations. This includes those crucial estimates of precision encoding the brain's certainty and uncertainty about external evidence and its own predictions. It is these latter estimates that provide much of the unhelpful "wiggle room" that seems to be exploited by the depressive brain, amplifying negative routines and delivering immunization against new positive evidence. This is the same mechanism thought to underlie some cases involving functional disorders in which symptoms manage to persist despite presenting the sufferer with good and apparently disconfirming evidence.

In all these cases, predictive processing offers a new and quite specific account of the neural organization involved. It is at this point that the psychiatric—even when it is clearly rooted in lifetime experiences—is revealed as neurological. This is also what makes the new theories testable—for example, by using neuroimaging techniques such as EEG to look at responses to social prediction errors (prediction errors arising in regard to social situations) in depressed versus nondepressed people, so as to discover whether some of these responses are indeed amplified in depressed individuals. Predictive processing here positions itself as a new and promising way of making sense of the complex interactions between brains, bodies, and social environments. If it is on the right track, depression is never simply a disorder of mood. It is a disorder of the whole body-brain-environment system, affecting the way the brain forms and runs bedrock bodily energy budgets, and the way it responds to new positive and negative information.

Aesthetic Chills (They're Multiplyin')

As one last—and deliberately very different—example of the rich interplay between bodily, emotional, and predictive factors, consider the "aesthetic chill." Also known as psychogenic chills, these are the distinctive shivery feelings (often paired with piloerection or goose bumps) that many people experience at moments of sudden high emotion. Aesthetic chills occur in many contexts, including as a response to art, film, poetry, scientific insight, social ritual, or even when watching a skilled sports performance. Importantly, the same kinds of chill response occur at less happy moments too, when we encounter fear or danger—such as when watching a horror movie.

According to what has become known as the "salience detection hypothesis," these chills occur when we encounter something that our brain identifies as critical new information that resolves important uncertainties. This makes it a kind of physiological echo of the "aha" moment when things suddenly fall into place. This is a moment at which an important pattern is first spotted, enabling us to predict the future in a new and powerful way. Music is one of the most reliable causes of aesthetic chills. This has made the induction of music-related chills a favorite tool for use in controlled experimental settings, including many neuroimaging investigations. This is because music is a domain in which expectancies and uncertainties are first generated at many levels, and then resolved at key musical moments.

This can seem paradoxical—why do we feel the chill that (in effect) says we have resolved a whole lot of dangling uncertainty when listening to a piece of music, or viewing a piece of theater, that we have experienced many times in the past? A full treatment of this would take us too far afield, but the key idea is that the power of great music (and great literature) lies in its ability to lead us through a staged process that first reliably builds up and then reliably resolves expectations. Recent work explores this idea in impressive detail, using neuroimaging and other techniques.

In the case of truly great works of music or literature, repeated encounters can also reveal new, deeper patterns. Importantly, however, the process of salient error reduction can occur even when we already know exactly how it is all going to end. This is quite closely akin to experiencing a familiar roller-coaster ride. We may have done that same ride a hundred times before, but the experience is carefully designed in ways that automatically engage our expectations, building them up and then delivering the satisfying resolution—time and time again! This is because they are—to use University of Oslo Professor of Comparative Literature Karin Kukkonen's memorable phrase—"probability designs": artifacts engineered to interact in reliable ways with our own predictive brains. Books, novels, plays, and movies are all probability designs. Attention (precision-weighting) plausibly plays a key role here, adding impact to the musical items that portend key moments in the movement or symphony. Aesthetic chills are a physiological marker of this sudden increase in estimated importance (precision).

Aesthetic chills thus provide further evidence (as if any was needed) of the deep two-way influence binding bodily and emotional response. This suggests the possibility of a little reverse engineering—of using the physical signature to enhance our own emotional response. The Fluid Interfaces Group at the MIT Media Lab have done just this, artificially inducing aesthetic chills using a "Frisson prosthesis." The device (see Fig. 4.1) includes three thermal way stations or "Peltier elements," a control board, and a vibrotactile unit. In operation, this setup delivers a wave of coldness that travels down the back of the wearer, imitating the unfolding of an affective chill. Artificially inducing the stream of sensation should nudge the brain into thinking that there has been a sudden resolution of some important and emotionally salient uncertainty. This would impact the way the person experiences whatever else is going on at that time, potentially enhancing the felt intensity of emotions. You might experience a moment in a play or musical movement as somehow even more salient and important as a result of the sudden artificially induced chill.

Preliminary results showed effects on reported emotional intensity. This work remains extremely exploratory, but the reverse engineering principle is sound. It resembles the "facial feedback hypothesis"—if you suddenly find your facial mus-

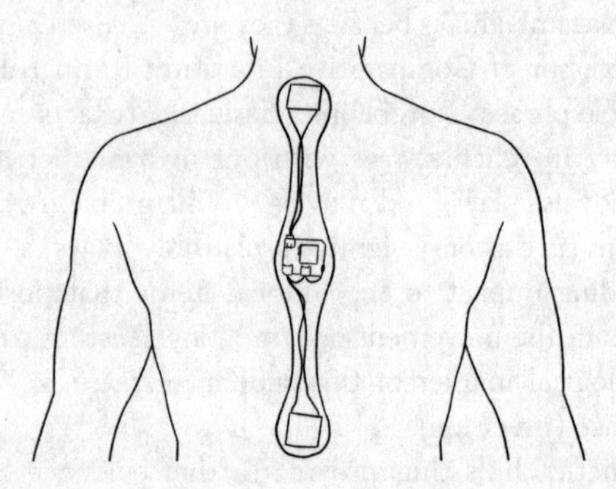

Fig. 4.1 The Frisson prosthesis: a device delivering thermal feedback in a manner closely resembling the affective chill

cles formed into the kinds of configuration they are normally in when you are happy (smiling) then that itself acts as a little bit of evidence, nudging the predictive brain toward the guess that you are feeling pretty good. In that way the very act of smiling (even if it is artificially induced, say by placing a pencil in the mouth) can potentially contribute some small amount to an actual experience of happiness. The Frisson prosthesis acts in the same basic way, artificially inducing a bodily sensation that the brain takes as evidence for the presence of its normal cause, namely a sudden and unexpected reduction of uncertainty.

•

Predictive brains look inward as well as outward, and it is those inward-looking dimensions (or so I have argued) that allow human experience to be constantly infused by feeling and emotion. This is because our take on the outside world (the way things look, taste, feel, and sound) is in constant two-way communication with information and predictions about our own changing internal physiological states. When this all works in harmony, it keeps us from straying too far from our window of bodily viability, and proactively budgets for our basic bodily needs. But when this system malfunctions and misregulates, it can lead to depression, anxiety, and retreat from the world.

Bodily prediction helps sculpt an experiential world in which some states and events are simply more attractive (hence more likely to be occupied) than others. This enables living beings to bring forth meaning and mattering from an otherwise meaningless material flux. We find ourselves drawn toward food and good company. Of course, there are circumstances in which we may actively seek out very different states, perhaps temporarily eliciting hunger and solitude as part of a fasting retreat. This speaks to the many time scales at which we may be trying to minimize prediction error relative to our goals—a complex issue to which we later (Chapter 6) return.

It is the ability to crunch together inner- and outward-looking sensory information that makes predictive brains such a valuable and life-preserving adaptive asset. But that praiseworthy tendency also has another, darker side. It allows distortive bodily information to sometimes tip the scales, unhelpfully impacting what we seem to see, hear, and feel. We have met some such cases already, both in the laboratory and in the origination of various delusions and hallucinations. But this tendency runs deep and has social and political consequences, as we will next see.

Interlude:

The Hard Problem—Predicting the Predictors?

WE HAVE seen the profound effects of differing expectations on all that we seem to see, hear, and feel. We have seen how variations in precision-weighted balancing acts (that determine the relative influence of sensory evidence and predictions) lead to variations in experience and how they lead to action too. And we have explored the origins of emotion, valence, and the sense of mattering, linking these to predictively valuable information concerning our own changing physiological states, and to the varying "error dynamics" that track how well or badly we are doing relative to our own expectations. This is a scientifically well-grounded account that provides powerful explanations for a wide swath of human experience.

But despite all that, you might feel that something is missing. What, you may ask, does all this tell us about the origins of what philosophers call "qualia"—the distinctive qualitative feel of "seeing red," "feeling angry," or "tasting like raspberry Kombucha"? Are all those exhilarating and nauseating qualitative experiences that populate our day-to-day mental life nothing over and above that multifaceted precision-weighted prediction machine in action?

I think you'd be right to take that dizzying step. But you might also feel a little bit wary. What seems to be missing, to put it bluntly, is an account of why our prediction-driven experiences feel the exact way that they do (indeed, why they feel any way at all). For surely—you might say—we could build a machine that makes extensive use of inward- and outward-looking prediction, that is sensitive to its own error dynamics, and that feels nothing at all? So, what is it about us embodied prediction engines that ushers all that "real feeling"—all those pesky qualia—onto the mental scene?

This, of course, is nothing other than the "hard problem of consciousness"—the one that has long been thought to represent the largest stumbling block for scientific attempts to explain the mind. In this highly speculative Interlude, I want to point toward a somewhat surprising answer. The hard problem we seem to confront is, in some ways, a trick of the mind. Don't get me wrong: it's not my view that conscious experience doesn't exist. But the key stumbling blocks are more conceptual than scientific.

An important clue lies in the subtle role of our own self-expectations in the predictive mix, and the way those self-expectations then become entangled with expectations about the wider world. It's these hidden entanglements, coupled (I'll argue) with our own unusually powerful cognitive resources, that then trick our predictive minds. The hope is that by understanding the factors that thus conspire to make our own conscious experience *seem* so very puzzling, we can begin to deflate the hard problem itself.

To approach the topic in this way is to start by tackling instead what David Chalmers has dubbed the "meta-problem of consciousness"—the problem of explaining why it is that we are drawn toward mind-body dualisms and tempted to posit an unbridgeable "explanatory gap" between the best science can offer and the facts about our own experience. Once that puzzle is solved, perhaps the hard problem itself will start to look different. Perhaps—though this is my hope more than

Chalmers's—we'll start to see that we have many of the right tools already in our hands.

This requires a note (perhaps a symphony) of caution. The various phenomena that underlie the hard problem are widely felt to be among the most important yet scientifically ill-understood features of the universe. Philosophers and scientists disagree, both with each other and among themselves, about what the problem really is, how best to solve it, or even if it is possible to solve it. The suggestions that follow won't appeal to everyone, and some will think they miss the point entirely. This is because they aim to alter our conception of just what it is that we really need to explain.

Simple Sentience

Work on the predictive brain already accounts for multiple interlocking features of lived human experience. It offers—we saw—a principled account of how inner and outer sensing work together to put us in touch with a structured world populated by meaningful possibilities for action. In addition, and deeply entangled with our grip on the outside world, an inward-looking (interoceptive) cycle targets our own changing physiological states—states involving the gut, viscera, blood pressure, heart rate, and the whole inner economy underlying hunger, thirst, and other bodily needs and appetites. Feelings and emotions then reflect predictions that integrate information involving interoceptive (bodily), proprioceptive (action-guiding), and exteroceptive (outward-looking) cues. For example, we just saw that interoceptive sensory information about current heart rate is used to help predict the presence or absence of visually (hence exteroceptively) perceived faces, so that neutral faces are more often seen as threatening when heart rate is increased.

As my bodily state alters, the salience of various worldly opportunities (to eat, for example) alters too. That means I will also act differently, harvesting different streams of information. Philosophers and psychologists talk here of "affordances," where these are the opportunities for action that arise when a certain type of creature encounters a certain kind of situation—a hungry green sea turtle encountering a nice patch of algae discovers an affordance for eating, whereas a human diver encounters a different affordance—perhaps it is an opportunity to photograph the turtle having its lunch. As our own bodily states alter, the salience (implemented by varying precision-weighting) of various worldly opportunities and affordances alters too. The sea turtle that has just eaten may not find the next patch of algae quite so attractive. The diver who has just captured the perfect turtle-lunch moment may now see a chance to explore the surrounding area in search of other creatures. Emotion—or so we argued—reflects the changing value of different actions given our bodily state, goals, needs, and projects. It is a kind of marker of our embodied attunement (or lack of it) to the world.

Moreover, as we also saw, much of the experienced valence of events and states of affairs (the way they present themselves to us as attractive or repellent, as ones to approach or to avoid) seem to reflect ongoing sensitivities to our own error dynamics. If our brains suddenly quash greater than expected amounts of prediction error—perhaps the diver now spots a very rare sea creature they had always wanted to film—we find ourselves "liking" the situation, and feel a strong urge to exploit it fully. In response, the brain increases its learning rate, amplifying the impact of that new salient information on the long-term model that guides action. The diver may then suddenly alter her plans for tomorrow, preparing to return to the same spot in the hope of a further sighting.

All this makes real progress with what might be thought

of as the problem of "basic sentience." The term sentience was used by the political and social philosopher Jeremy Bentham as long ago as 1789 (in his *Introduction to the Principles of Morals and Legislation*) to mark a distinction between the capacity to feel and the capacity to think and reason. Creatures lacking the full sweep of human capacities to think and reason, it was argued, might nonetheless be capable of feeling pain and pleasure, of appreciating how well or badly things are going, and of learning from their experiences.

We can now think of sentient beings as those whose neural model of the world is in constant two-way communication with a model of their own changing physiological state. Basic sentience emerges in creatures whose sensitivities to states of the external world are subtly but pervasively responsive to the likely future states of their own bodies and metabolisms. These creatures don't just see a tree, or a shadow—they see a source of much needed food, or the threat of an imminent attack. Such creatures will perceptually encounter a world fit for action, in which what actions are selected depends heavily upon a sense of their current and ongoing bodily state and needs, and how well they are doing at minimizing salient error. They live, we might say, in a world that is temporally extended and perceptually meaningful. Bodily self-regulation, action, and temporal depth are, predictive processing thus suggests, jointly necessary if there is to be conscious experience at all.

Creatures like that will certainly *appear* sentient. They will respond to their worlds in ways informed by a delicate dance between what they detect in the external world and their own ever-changing bodily needs and states. This, I argue, is what underlies all the behavioral manifestations of "sentience." We detect sentience in creatures (and potentially in robots) whose take on the external world is subtly but pervasively responsive to their brain or control system's take on their own inner, bodily worlds and their own states of action readiness.

Does this constitute true sentience, or might it still be merely apparent sentience? I am not convinced that this is the right question to ask. Let's just say (for the moment) that predictive processing offers a promising story about how behavioral patterns like that might be ushered into being by the constant interanimated effort to predict internal and external sensory variation, and to minimize prediction errors as we do so. We can then ask, what else could be going on when we humans go one step further and start spontaneously to report (and puzzle over) the presence, in our own experience, of all those distinctive looks, feels, and "qualia"?

Expecting Ourselves

The next step borrows a key move from the philosopher Daniel Dennett. Back in 2011, I spent some time marooned (thanks to Hurricane Irene) with Dan and some of his students in his atmospheric farmhouse at Blue Hill, Maine. The hurricane had left us without power, forced to make our own amusements in Dan's wind-and-rain-battered farmhouse. Discussion was unhurried but determined—we were going to solve the mind. A recurrent theme, one that kept nagging at me afterward, was just how predictive brains might behave when they turned their formidable resources doubly inward—not just upon their own bodies and error dynamics but also upon their own predictions, behaviors, and responses. Might this "predicting of our own predictions" hold one of the missing keys to understanding our own conscious experience? To see how it might, Dennett reminded us of the power of what he has long dubbed the "strange inversion."

Here's my way of introducing this important idea. Some bars are famous for their excellent Guinness. You might think that this is due to the expertise of the barkeeps, or something special about the plumbing. It turns out, however, that the major determinant of the goodness of Guinness is how long the barrel has been open. The shorter the length of time, the better the beer. This opens the door to a strange inversion in our understanding of the cause of the bar's reputation. We think that the bar's reputation is due to the unusual goodness of the Guinness it serves. But in fact, the bar's reputation for serving good Guinness is what makes so many people order Guinness there, and it is the resulting rapid turnover that makes the Guinness there taste so good.

Good for Guinness, you may say. But what can this tell us about the nature and possibility of conscious experience? Dennett asks us to conjure up the taste of honey—the specific experiential feel, the subjective taste of honey, the elusive "qualia" themselves. According to the standard story, we might say we like (or possibly loathe, it can work either way) that sweet taste. Our liking or loathing presents itself as a response to our own experience. But perhaps this gets the experiential cart before the behavioral horse, in roughly the same way as with the goodness of the Guinness.

Applied to the taste of the honey, Dennett's strange inversion works like this. The specific and elusive "taste of the honey" is nothing but the subtle complex of responses it happens to evoke in me—responses that include seeking it out, licking it off the dipper, spreading it over a biscuit, pronouncing it to be tasty, and so on. Tasting like honey is then simply the way I label things that predictably evoke, in me, that specific complex of (actual and possible) responses. In other words, the facts about my web of possible responses are not the result of the experienced taste. Instead, when that web is in place, that is what I intuitively call "the taste." The web of behaviors and responses comes first, and the puzzling, ineffable taste is really just a handy label for that web.

But advanced intelligences though we are, we don't automatically know this. Instead, we find ourselves able fluently to predict our own and others' responses by modeling ourselves using some simple shorthands that suffice for our daily purposes. That simple shorthand says we are home to "qualitative experiences," many of which we either like or dislike. Knowing that you like sweet things I can select a Drambuie Scotch liqueur for you this holiday season, rather than a peaty malt. Knowing that I love salty things rather too much helps me keep an eye on my otherwise boundless consumption of cheese. The shorthand self-model works. But if we are not careful, it can lead us toward a kind of metaphysical inflation. We then start to take that model very seriously indeed. We posit the existence of a strange experiential realm in need of some special, currently unimaginable, kind of scientific explanation.

If this is correct, then spotting ordinary things like instances of dogs and cats (which seem to just be "things in the world") and stranger things, like the "taste of honey," are all actually on a par—they are all just inferred causes that the brain conjures up to help us predict our own sensory flows. But in the latter case, a major part of what we are predicting is ourselves: our own matrix of hidden tendencies to act and respond—to lick, and to exclaim "oh how I love that Manuka honey." Such is the view from after the strange inversion.

You might worry that we can (of course) taste something brand-new or discover that we actually like a taste we just didn't expect to like. This is true but (or so I'd argue) really poses no extra puzzle. As we learn the taste of something new, we are learning to predict the multiple complex responses of our own taste receptors, as well as those of other sensory organs responding to the item's look, smell, and texture. We will also find ourselves reacting to it in some way—perhaps by approaching it, avoiding it, or sampling it again. This would support a version of the kind of model learning described in Chapter 1. At that point, the stage is already set for the strange inversion described above.

Once in command of a structured predictive model, we may quite easily spot some never-before-seen object as being a new kind of cute animal, sweet treat, car, or food mixer. We can do this because our own prior learning experience has in each case allowed the brain to lock on to a complex of subtle interlocking cues and features. These can later be spotted when they co-occur, even in brand-new instantiations. Thus, if some never-before-seen animal has soft fur, large eyes, and a disproportionally large head, then (other things being equal) it will immediately be recognized as falling under the existing label "cute." I might even be surprised that the new animal looks cute, having consciously expected it to be scary. But my predictive brain recognized it as another instance of the same set of subtle co-occurring features present in previous cases of cuteness. Similarly for a new food item that, even on first encounter, tastes pleasantly sweet, or surprisingly salty. When new inputs are swept under some existing predictive umbrella in this way, this will also tend to recruit my established behavioral tendencies, such as approaching things that look cute.

Simple Self-Models

When we perceive the world, we weave together information about what's out there with information about our own inner physiological states and our own tendencies to action. This is how animals of all kinds get to experience a structured world of opportunities for action and intervention, and it is (I believe) what makes them into sentient, feeling beings: beings that find themselves in a world where things really matter.

What would such beings say if asked just what they find and detect during their perceptual encounters with the world? Before addressing this issue, it is important to remind ourselves that the presence of the kind of organization I have just described is consistent with the complete absence of advanced (indeed, of any) capacities for verbal rehearsal and report. I am not assuming that experience occurs only in the presence of human-like language and advanced thought. Far from it. But we are now turning our attention to a new target: not bedrock sentience but our feelings of puzzlement about our own conscious experience.

We still (and quite badly) need a better understanding of just how and why some of the brain's best guesses become positioned to drive verbal report and other forms of report-like behavior—behavior that reveals what a person or animal currently (overall, all things considered) takes to be the case. But this is, in principle, just the kind of puzzle that the existing sciences of the mind (including predictive processing) should one day resolve. Given a certain visual input, a language-using sentient being might say, for example: "I detect a large and rather cute dog, probably some kind of Labrador cross, whose coat is reddish brown—perhaps it is a Labradoodle."

The point to note is that every property and feature here reported has—according to predictive processing—been extracted in exactly the same way. Each feature and property has simply been inferred as part of the current best attempt to predict the current waves of sensory stimulation. Redness, largeness, cuteness, dogness, and Labrador-ness all emerge as inferred causes, designed to support fluent prediction and action control. They all serve to organize and predict whole swaths of interacting inward- and outward-looking sensory information. Nonetheless, a sufficiently intelligent self-expecting agent will very soon start to be puzzled by what they may start to describe as the elusive "qualitative dimensions" of their own experiences.

Imagine a robot that can be quizzed about what it sees and about its own reasons for action. But imagine also that the robot, when forming its reports, has access only to how the predictive model running in its inner machinery currently estimates bodily and environmental states to be, rather than to the full details of the processing that leads to those estimations. This makes good design sense after all. The whole point of the process of probabilistic inference running inside its silicon brain is (let's assume) to estimate how things are in the twin arenas—its own body and the world—relevant to its own survival and to successful action. Knowledge that looks computationally inward, at all the intervening details of the processing stream itself, would be adaptively redundant. Worse still, it would incur metabolic cost without any clear benefit.

We know this firsthand, since when we humans see the world we do not experience the many steps of our own visual processing. We are blithely unaware of the computations being performed by different neuronal populations in early visual processing areas. All that nature cared about when building us was that we be able to detect what's out there (is that a predator or a friend?) and how things are physiologically, whether we need food, water, or rest, or to avoid injury. Adding machinery to enable us to appreciate the full complexities of our own inner processing would have been costly and quite possibly counterproductive, diverting attention from what really matters in body and world.

A being with that kind of genuine but limited access to their own processing will experience a world populated by dogs, cats, chairs, hurricanes—but also (so I'd argue) by "scary movies" and "drinks that taste like raspberry Kombucha." These are all neural best guesses about what's out there and in what ways it matters to an embodied organism with time horizons and metabolic needs. But unable to appreciate our own inner predictive regimes, we make do with a highly simplified picture of ourselves as simply "seeing dogs," "feeling pains," "feeling hungry," etc.

This is how we "predict the predictors"—both ourselves and others like us. But these models (our own predictive pictures of ourselves) are under pressure to be as simple as they can possibly be, capturing only what is needed to support behavioral and adaptive success.

Questioning the Philosophical Zombie

The intelligent agent, armed only with these stripped-down, efficient, self-predictive models, finds themselves in very much the situation once described by Chalmers himself. Back in 1996, Chalmers asked himself how the process of perception would strike an advanced intelligence that has access only to the end products (the best guesses, as we'd say) resulting from their own complex inner processing. His reply was that such an agent, when asked how they know that the honey tastes sweet, may be forced to say they know this directly, in some brute but puzzling manner. They might then start to judge that they are home to mysterious "qualia."

Chalmers himself does not think this move can ultimately deliver the goods. Specifically, he then wonders why the agent's experience strikes them as having a perceptual quality at all—why is it not like "just knowing what is there" without any perceptual character whatsoever? This would correspond to a kind of "darkness within"—a perceptual zombie system that can output "look at that lovely cuddly puppy with the bright brown hair" in the presence of the lovely cuddly puppy but without any accompanying perceptual or emotional experience. Chalmers himself has frequently invoked the possibility of full "philosophical zombies" in arguments concerning consciousness, where a full zombie would be a creature all of whose behaviors (including everything they say and do) perfectly match our own, but one that is lacking any form of inner mental life or subjectivity.

It is true that when approaching these tricky issues, we need to be careful not to build the problematic notion of qualitative experience into the very idea of a creature's sensory best guessing. Instead, we should be using a notion of predictions and best guesses that is "experientially neutral." It is meant to be agnostic, that is to say, about whether the best guessing emerges with or without any accompanying feelings or sensory experience. After all, a clever algorithm can caption an image, spotting that it contains a dog (say) without experiencing anything at all.

But the more detail we plug in here, the less plausible the "full zombie" picture really seems. Our imagined being will know a lot more than simply "that there is a dog (say) out there." They will know the shape, color, texture, and behavioral tendencies of the animal. They will know (roughly at least) that some of their own knowledge comes from sight, and other parts from hearing and touch. They can learn how their sensory best guesses would alter and fluctuate were they to cover their eyes or ears. They will know that typical dogs have four legs and one tail and will be able actively to seek out additional visual evidence for each such feature by attending more closely (upping the precision on) certain spatial regions. Most importantly perhaps, they may know that they generally "like" dogs—they tend to respond positively to their looks and seek out opportunities to pet and play with them.

Could all that happen "in the dark" experientially speaking? The more you examine such a claim, the harder it becomes to really imagine it. The being just described sounds a whole lot like us. Such a being does not know how their knowledge about the world actually comes about. They do not know that that process merges bodily and exteroceptive evidence. Nor do they know that their own picture of themselves is drawn using an effective but highly simplified schema in which they simply "see dogs" and "like dogs." They are thus poised to find their own subjective experience strange and puzzling in just the ways Chalmers described.

I believe that we are those creatures. Self-opacity and simplified self-models lead us—clever but limited beings—to infer that we are home to a mass of extremely puzzling "qualitative states." But these inferred qualia are just another handy tool for predicting ourselves and others. The underlying states (sentient beings' best guesses at how things are in body and world) are real. But our profound metaphysical puzzlement is mistaken.

Qualitative consciousness is real. But perhaps (just perhaps) it isn't exactly what we think.

5

EXPECTING BETTER

IT WAS the Dutch microscopist Antonie van Leeuwenhoek who, in 1677, first saw spermatozoa under the microscope. Leeuwenhoek was, however, already something of a convert to "preformationism"—the idea that adult bodies are fully but minutely present in human sperm. Reporting his experience, he claims to have actually seen in the semen "all manner of great and small vessels, so various and so numerous that I do not doubt that they be nerves, arteries and veins. . . . And when I saw them, I felt convinced that, in no full-grown body, are there any vessels which may not be found likewise in semen." His visual experiences here seem to bear out his strong prior beliefs. But looking back on this today, we may well suspect that it was really the other way around—that Leeuwenhoek's visual experience reflected nothing so much as those strong prior beliefs.

The story about Leeuwenhoek may or may not be historically accurate. But it dramatizes a kind of "wishful seeing" that many of us may have experienced at some time during our lives. Predictive processing makes sense of such cases. It suggests that the way we see and experience the world is quite routinely shaped and guided by our own (often unconscious) predictions and expectations.

This is a huge asset. It enables us to tell how things are out there in the world even when available sensory information is impoverished or ambiguous. It helps us spot the predator well hidden in the bushes, or the signs of cancer barely visible on a fuzzy X-ray. But there is a darker side to all that disambiguating prediction too. For wherever prediction helps construct experience there is a kind of bias. The world as we see and sense it becomes shaped, in part, by our own (conscious and unconscious) expectations. This is not merely bias in thought or judgment but bias affecting the primary sensory realm—the source of our apparent evidence—itself.

This chapter pursues some of the ways predictive brains can cause us to make such mistakes, leading us to misperceive people, events, and objects in the world. It also asks what can be done to combat our own predictively biased perceptions.

Perceiving What You Feel

A simple example, lightly adapted from work by the Harvard philosopher Susanna Siegel, can help set the scene. Imagine that you believe that your friend Jack is angry with you. When you see Jack, your strong belief that he is angry sculpts your experience in just the ways we have seen in previous chapters. Your friend's face, which to most observers would currently look neutral, now seems to you to display small but telling visual signs of anger. You may think you have now garnered some useful evidence for your prior belief that Jack is angry. But—as Siegel points out—there seems to be something rather odd about this, since that subtle "perceptual evidence" is itself a manifestation, in perception, of that very belief. The belief itself is what tipped the scales, delivering a visual experience (of Angry Jack) that now seems to lend support to that same belief (that Jack is angry). This looks, Siegel argues, like "double-counting."

Consider also the likely practical consequences of such a

mistaken visual experience. One likely consequence is that you may start to act a little bit differently. This, in turn, will be visible to your friend, who may begin to behave oddly toward you. This genuine difference in behavior might seem to you to provide still further evidence that something is indeed amiss. As this cycle repeats, you might both become a little angry, each unable to see where, and why, things went wrong.

Perceptual effects of the Angry Jack kind, as we saw earlier, have been shown in controlled laboratory settings. In some of those experiments, false cardiac feedback led participants to perceive visually neutral faces as more emotionally intense. In Siegel's example, the prior belief that Jack is angry plays a similar role to that of the additional cardiac evidence in these experiments. But these issues matter outside the laboratory too. "Shooter bias" among police officers provides a chilling example. A recent paper looking at shooter bias noted that between 2007 and 2014 a full 49 percent of officer-involved shootings of unarmed victims were linked to what have become known as "threat perception failures." In such cases, officers mistook an innocent object such as a cell phone, or a nonthreatening movement, as an armed threat. Moreover—as has become increasingly appreciated in recent years—these threat perception failures predominantly involved Black victims. This was true in 80 percent of such cases in the city of Philadelphia where the numbers of White and Black citizens are roughly equal.

In seminal work on this topic, Professor Lisa Feldman Barrett describes what she dubs "affective realism effects" whereby a police officer's own inner bodily sensations (for example, heightened heart rate, sweaty palms, clenched stomach, facial flush) may be taken by the predictive brain as additional "evidence" for active threat, leading them to perceptually misidentify innocent items, such as a smartphone, as drawn weapons—especially when those items are held by

Black men. These are, of course, the very same deep involvements of bodily signals with neural predictions celebrated, in the previous chapter, as helping to create the feelings of mattering that give our mental life so much of its shape and flavor. Here, though, they are also a contributing factor to tragic gun-hallucinations.

The bodily sensations themselves, Barrett notes, might have many causes, including long shifts, previous encounters, and even the ingestion of caffeine. They will also be caused by a heightened sense of danger, perhaps prompted by darkness or location. But acting in concert with misguided racial stereotypes, these bodily cues will sculpt the predictions about what's out there that deliver visual experience. Similarly, an individual who has just experienced anger (induced in a controlled experimental setting) has been shown to be more prone to identify an innocent object as a gun than those who had been caused to feel a different prior emotion, such as sadness. The upshot is that, to use an evocative phrase from this literature, you will sometimes "perceive what you feel." This is just another manifestation of that continuous line linking bodily signals, emotion, and our perceptual experiences of the wider world.

Responding to Predictive Bias

How should we respond to these alarming demonstrations? They should certainly not lead us to absolve police officers from responsibility for these kinds of awful mistakes—clearly, racist stereotypes play a major role in such scenarios. There are also many ways in which larger-scale police culture is at fault, since this has allowed practices and expectations to take root that reflect a kind of collective bias. These practices and expectations lead to escalating cycles of conflict. Individuals, perhaps especially those working on the front lines, also have a responsibility to educate themselves about the effects of misleading cultural stereotypes. With improved understanding comes new room for hope. If it is indeed the combination of misplaced stereotypes with distortive bodily information that delivers some of the most dangerous hallucinations, then we can leverage change not just by challenging bad stereotypes but perhaps also (more on this later) by using new training regimes to alter our responses to that bodily information.

Understanding how misplaced stereotypes and unjust practices interact with our own changing bodily signals should add to the growing realization that what is most urgently needed is deep and abiding change in bedrock societal practices and institutions. These include media depictions, police culture, and press coverage, all of which play a role in encouraging the conscious and unconscious racist beliefs that lead to misguided predictions: predictions that distort perceptions and result in inflammatory and sometimes fatal interactions.

The influence of media depictions is powerfully illustrated by another set of studies conducted by Barrett and colleagues in 2013, just one month after the Boston Marathon bombing. Participants were shown identical images paired with headlines taken from newspapers around that time. Some of the headlines highlighted threat ("Not Since 9/11") while others reported community spirit and healing ("Boston Strong"). Tested afterward on a "threat perception task" in which the goal was to shoot at simulated armed targets and avoid shooting at unarmed ones, those exposed to the threat-framed stimuli made more misidentification errors, shooting at unarmed targets, than those exposed to the more positively framed versions. Their brains were, in effect, predicting new threats rather than better futures. This is not news. It is well known that the many ways we structure our world, including the way we express things in newspaper headlines, subtly alters our own subsequent perceptions and responses. In the case of the news headlines, these effects were amplified in subjects who reported that they had been very strongly affected at the time by the bombings as compared to those who rated their original emotional response less strongly. This makes sense since strong emotions, in predictive processing, are associated with high estimated precision—so the brain is treating that information as unusually significant and reliable. In all these interlocking ways, the authors of the study conclude, "feeling significantly distressed or threatened can predictively contribute to perceiving the world as more stressful or threatening in a very literal sense."

Helpful Fictions

One place we can clearly intervene is by altering the environments that train our own prediction machinery in the first place. Obviously, the gold standard here would be to build worlds in which racist and sexist patterns, either real or depicted, were never encountered. But long before that golden day arrives, small tweaks and changes can be effective too.

To take just one example, there is currently, in the U.K., a strong association between being a plumber and being male. This is a real pattern that well-functioning predictive brains cannot help but detect. Similarly, figures from the U.S. in 2012 showed that 80 percent of engineering recruits were male. In cases such as these, it is the cultural milieu generating those real statistics that needs to be altered. But this is a long and difficult process.

Fictional worlds provide one small, much more easily manipulated, lever for change. We humans do much of our learning in media, advertising, and entertainment. We read books and see movies, we play video games, some of which may involve immersive virtual realities combining passive perception and real-world action. This is both a vulnerability and an opportunity. It is a vulnerability because many of our constructed fictional worlds either fail to reflect the true nature of our societies or reflect aspects that we would not wish to promote. They may depict misinformed or unhelpful racial and gender role stereotypes, or highly unrepresentative body shapes. But there is an opportunity here too. We can act to remedy this, simply by proactively structuring more of our fictional worlds in ways that are more realistic, or that are helpfully aspirational. In so doing we can begin to push back against racial, bodily, and gender role stereotyping.

Fictional worlds can play a unique and valuable role both in installing new and better expectations, and in challenging old ones. Importantly, many of the elements in need of changing may not have made it into our conscious awareness or become objects for our own critical attention. By retraining our unconscious prediction machinery, fictional worlds can act as powerful tools for pushing back here too. They can reduce estimated certainty regarding existing pernicious stereotypes and help install new and better ones

Immersive virtual reality provides what could prove to be the most potent of all such interventions. Like video game playing, immersive VR allows for agency and action, and agency and action are prime ways of training the predictive brain. An example here is the role of virtual bodies as part of a new and effective intervention for anorexia nervosa. Participants with anorexia were first encouraged to experience a virtual body with a healthy BMI (body mass index) as if it were their own. To encourage this, subjects used a VR headset to view the abdomen of their healthy BMI VR body being touched and stroked with a soft brush while simultaneously feeling an identical touching and stroking routine applied to their real abdomen. This encouraged them to identify fully and viscerally with the VR body. Later, when asked to estimate the size of their own (nonvirtual) body parts, those trained with the healthy BMI virtual body showed a reduced tendency to overestimate the size of their actual body parts—height; shoulder width; abdomen width; hip width; shoulder circumference; abdomen circumference; and hip circumference.

Immersive virtual realities also show great promise as new arenas for police training. Crucially, VR enables police officers to practice (with informative feedback) the kinds of rapid decision making and action taking that will characterize real-world encounters. Getting action safely into the training circuit is essential since real-world visual experiences are often constructed under time pressure and as part of an action-perception loop.

Improving Interoception

Other forms of training could directly target some of the internal bodily signals at work in cases of "seeing what you feel." New training protocols can help us become more aware of our own shifting states of physiological arousal. This is already being achieved in a few trial programs using simple wearable devices to deliver ongoing biofeedback information—information concerning key parameters such as heart rate. Our own cardiac signals, as we saw, can easily mislead the brain's perceptual inference engines in high-stress situations. But systematic training regimes can teach ways to reduce this arousal, even during real, potentially dangerous, engagements. Careful bodily awareness training of these kinds might help avoid at least some lethal force errors.

Such training, as Barrett notes, might even deliver health benefits for the officers themselves, who are known to be at increased risk of heart disease, PTSD, and obesity. Interoceptive sensitivity also varies between individuals, as do tendencies to notice our own physiological states. Selection and training procedures could one day take such baseline differences into account, perhaps delivering more personalized officer-training regimes as a result.

Interoceptive training regimes are already being explored as treatments for anxiety. For example, my one-time University of Sussex colleague Professor Sarah Garfinkel has been exploring interventions that improve cardiac self-awareness as a treatment for various forms of anxiety. She found that people with anxiety are often very internally focused, while at the same time surprisingly bad at knowing their own heartbeat. In other words, they focus hard on their own internal state but do so without much accuracy or precision. Importantly, Garfinkel found that anxiety was most strongly associated with the combination of low accuracy regarding your own internal state and an inflated sense of that accuracy. This means that you are more likely to suffer anxiety if you are interoceptively inaccurate and yet falsely believe yourself to be very accurate.

Here too, predictive processing accounts deliver a neat diagnosis of the experimental finding. For it is that specific combination (inaccurate interoception with high self-estimated accuracy) that is most likely to deliver misleading inferences about what's happening to you, or what's happening in the wider world. Inaccurate and "coarse" information that is wrongly estimated to be both informative and precise will rapidly lead predictive minds to confident but misguided conclusions. By improving our own interoceptive accuracy by means of biofeedback training, it may be possible to minimize or avoid these kinds of mistaken inference, allowing individuals to contextualize their own bodily responses in more helpful ways.

Garfinkel's latest work seems to show just that, finding reduced anxiety as interoceptive accuracy increases. An extreme example of such accuracy emerged while she was working with a leading hostage negotiator—their interoceptive self-accuracy on the heartbeat task was 100 percent. Garfinkel speculates that this extreme self-accuracy plays some yet-to-be-fully-understood role in the hostage negotiator's ability to pick up "empathically" on how others are feeling so as to judge when and how best to intervene. Other work by Garfinkel and colleagues has shown what seems to be a related effect in individuals with autism spectrum condition. In this work, participants with better interoceptive self-awareness were also better able to detect the emotional information "hidden" in other people's speech intonations. This suggests that training that improves interoceptive self-awareness in people with autism spectrum condition may thereby improve their ability to discern subtle emotional information.

Closer to the Truth?

We have focused on some clear (and often tragic) cases in which predictions mislead, causing us to see or hear things that are not really there. But this leaves open a thorny, if rather more philosophical, problem. How much should predictions and prior knowledge contribute to perception to reveal things "as they really are?" The question is trickier than it sounds.

Consider puzzling images (such as the Mooney images seen in Chapter 1) that suddenly make visual sense after you have seen the original, or that make sense only after considerable effort. Another example of the latter kind is shown in Fig. 5.1.

Once you have managed to see the large cow face in the top left part of the image (look for the two black ears facing you, and the nose near the bottom left) there is no going back. You cannot then see the image the way you first did, as just a pattern of dark and light patches. "Cannot Unsee" is the internet meme that sometimes reflects this striking fact. For example, Fig. 5.2 is the famous logo from the 2014 FIFA World Cup.

In a tweet that was retweeted endlessly during the competition, a copywriter named Holly Brockwell wrote:

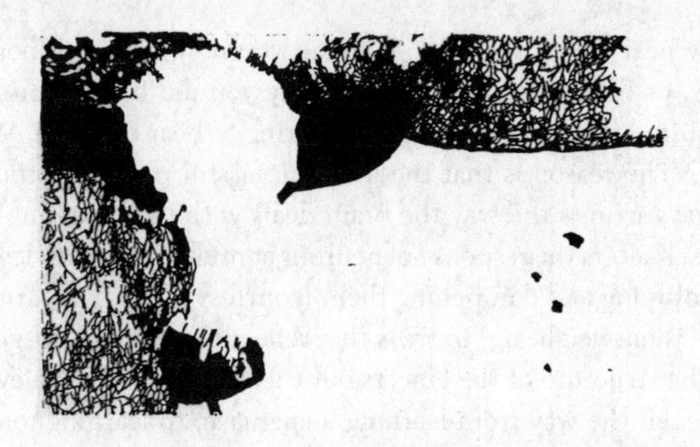

Fig. 5.1 There is an image hidden in the visual noise.

CANNOT UNSEE: the Brazil 2014 logo has been criticised for "looking like a facepalm."

The Cannot Unsee meme flags something interesting. Once we have a potent predictive model in place, we cannot usually undo it to reexperience the world the way we did before. Before you learned to speak your native tongue, what did utterances in that language sound like to you? You cannot

Fig. 5.2 Cannot Unsee (Facepalm)

now hear things that way. Nor can you now see the Mooney, cow, or FIFA logo images in the way you did before you had acquired the knowledge you now bring to bear on them. Why not? The reason is that the most successful predictive model always sculpts the way the brain deals with the incoming signals. It alters the response of neurons at multiple cortical levels, amplifying and dampening them (courtesy of all that variable precision-weighting) in ways that reflect the brain's best guess at the structure of the objects out there in the world. In every case, all the way from learning a language, to learning to distinguish a poplar from an aspen, or to spotting signs of cancer in fuzzy X-rays, our perceptual experience shifts and alters as new or better predictive models are formed, recruiting different precision-weighting regimes.

Precision-weighted prediction (usually) serves our purposes by highlighting some things at the expense of others. This is just the standard operating procedure of the predictive brain. As we learn more about our worlds, model-based predictions play a larger and larger role in sculpting our experience. But this leads to a worry. Are those predictions in some way acting like a veil separating us from the true nature of "the world out there"? Were your percepts perhaps closer to the truth when you were a baby, *before* you brought so much predictive knowledge to bear?

That cannot be right in any practical sense. To be sure, utterances in your native tongue must have sounded like something before you spoke the language, and something quite unlike the way you currently experience them. But there too, your brain was making guesses based on what it knew. It is just that what it knows now is different from what it knew then. More importantly, knowing the language enables you to spot things that really are there—the sequences of spoken words, each with a distinct identity and meaning.

Back in Chapter 1 we met the compelling example of

"sine-wave speech." This was ordinary speech stripped down to a simple sonic skeleton. On first exposure, it sounds like a sequence of meaningless beeps and whistles. But once you have heard the original sentence or have heard enough samples to become a fluent hearer of "sine language," your auditory experience is transformed, and the words and sentences uttered shine powerfully through. What was once beeps and whistles emerges as clear and meaningful speech. But when did you hear what was *really* there? Was it the first time around, when you very clearly heard the rising and falling tones, but before new skills and prior knowledge conspired to enable you to distinguish the various words? Or was it only later, when you could more readily identify the spoken, meaningful words?

First time around you were probably sensitive to more of the actual sonic waveform hitting your ears—for example, you will have heard the sound correctly as being a continuous stream. That information gets somewhat obscured later, when we seem to hear gaps between the words despite the continuity of the underlying sound stream. That's also why, when we listen to a spoken language we do not know, the speech can often sound unusually fast—our brain is not assigning boundaries in the ways that create those apparent gaps. But in such cases we are also failing to sift and shape sonic information in the ways that best reveal what matters most about the true original source—a specific string of spoken words.

Which of these "ways of hearing" is closer to the truth? It depends on what you are trying to do. Are you a sound engineer trying to detect something acoustically odd about a room? If so, then you will give certain aspects of the acoustic evidence extra weight as you try to track down the problem, attending to different possibilities in turn. Or are you at a busy party trying to hear what's being said against a noisy backdrop? Every scenario requires a different set of discriminations. In predictive processing terms, that means deploying a different set of active predictions and associated precision-weightings. All this suggests that we can never simply experience "the way things really are," or the "true signal from the world." Indeed, if predictive processing is a good account of perception it is not even clear what that could mean. To perceive is to bring (weighted) predictions to bear on the incoming sensory signals, and experience arises as these twin elements meet.

That does not mean we can never get things wrong. But it does mean that there is no single way of getting things right. To return to a metaphor used in Chapter 1, perception is more like painting than some kind of point-and-shoot photography—it is an act of creation that draws upon our own needs and history. In this act of creation, there can be no such thing as a perfect rendition of the raw incoming signal. Instead, we bring ourselves (our past experience and our current projects) to bear on incoming sensory signals. Predictions, anticipating the future and permeated with the past, shape human experience in all its forms.

The Active Keyboard

This means that in order to experience a world at all, we must always in some way be bringing *ourselves* to bear on that world. Back in the mid-twentieth century, the French phenomenologist Maurice Merleau-Ponty tried to capture this sense of active involvement using the image of a mobile keyboard—one that moved itself around offering up different keys up the monotonous beat of an "external hammer." The hammer was the world, but the variety of human experience reflected the different keys being offered up to that same beating hammer. The experienced world was like the message typed onto the keyboard: a message that in the end said as much about the action of the keyboard as anything else.

Predictive processing helps make sense of this (initially

quite puzzling) imagery. The most basic way that we actively construct our world is by selective sampling. We move our body and aim our gaze in ways that reflect what we expect to encounter. In this way, different kinds of animals, and humans with different individual histories, will harvest different sets of stimulations from the very same world. But as we selectively harvest those stimulations, our brains impose structure a second time, processing the sensory information in ways that amplify and dampen, extracting meaningful structure that itself reflects our own prior experience. The "predictive keyboard" is thus not just an active selector, but also an active processor of whatever gets selected.

In the case of sine-wave speech as it was experienced before and after learning to hear the hidden sentences, each balance reveals different aspects of what is really there. We should say the same about many neurodiverse ways of experiencing the world. In autism spectrum condition, for example, we saw back in Chapter 2 that there seems to be a much greater emphasis on the detail of incoming sensory information, with less dampening due to prediction and expectation. Is that closer to, or further from, the truth? The answer is neither. This matters more generally too, since neurodiversity is everywhere. Individuals operating within the more "typical" ranges will still display subtly different tendencies to weight sensory information against top-down guessing, or (as we saw in the case of PTSD in Chapter 2 and depression in Chapter 4) to update their predictions when new information becomes available.

We cannot help but base our current waves of prediction on our own native tendencies and particular life histories. Where those predictions vary, so does human experience. My University of Sussex colleague Professor Anil Seth sums this up quoting Anaïs Nin, "We do not see things as they are, we see them as we are."

Keeping It Real

Nonetheless, multiple factors work to keep us more or less grounded in a shared take on reality. One such factor is simply the presence of a mind-independent world and a human-specific suite of mechanisms for sensing and bodily action. Our physical world itself exerts an admirable tendency to resist wildly mistaken guesses at its shape and powers. Square pegs simply do not fit into round holes. Our early guesses at understanding things get refined and altered until they work. Some of what works best may become compressed into fast, efficient linkages that give us an ultra-rapid heads-up on the gist of the scene before us. There are also many shared features that characterize human bodies, brains, and nervous systems. Such common structures and pathways (including the gross anatomy of the brain) must play a key anchoring role in the construction of a shared human reality. So human experience reflects a shifting amalgam of these deep-set structural constraints and flexible, higher-level, "top-down" influences of the kind we have mostly been considering.

As social and highly communicative beings, we are also driven to try to bring our individually diverse models and expectations into line with each other—enough into line, at any rate, to facilitate talk, commerce, and social exchange. Those predictive models are also constantly tested by action. If my prior beliefs lead me at first glance to see the vague shape in the garden as the outline of my dog, Fido, my next action might be to look for Fido's distinctive pointy tail. If I don't find it, my first guess is rejected. Perhaps it is not Fido but a fox? That new guess sends me searching for different confirming features, such as an especially furtive gait. This shows that we shouldn't focus too heavily upon isolated "snapshot moments" in thinking about the veridicality (or otherwise) of perception but should instead look at how current best guesses respond to exploratory actions designed to test them. "Test then flexibly update" is usually a good recipe for long-term success.

•

Expectations, many of them unconscious, are always at work as our brains construct our experiences. Such effects are inevitable and can be extremely helpful. Suppose it's been a while since a loved one posted anything on social media. That small piece of new evidence might cause me, when next I see them, to attend differently, and so to spot the very faintest expression of worry in their face. That small piece of additional evidence (their social media absence) had caused me to increase the precision-weighting on very small facial clues, allowing me to spot a genuine sign of distress that I would otherwise have missed.

That's not "wishful seeing." It's seeing what's really there, but having it become visible thanks in part to the action of my own prior knowledge and expectations. Nor is this pernicious double-counting. Rather, it shows the operation of a predictive mechanism that is efficient and effective in most ordinary situations. Yet it is that same mechanism that is at work when things go wrong—when we hallucinate Bing Crosby singing "White Christmas," or seem to see a neutral face as subtly threatening thanks to bodily signals from our own fast-beating heart. Such effects, especially in time-pressured real-world situations, can lead to terrible consequences.

To begin to combat this, the first crucial step is simply better to understand all the possible contributing factors. In the case of shooter bias, this means understanding how high bodily arousal can add invisible fuel to the fires already lit by racially biased expectations. Understanding the role of prediction and bodily signals in this process should now be mandatory training for those in many (perhaps all) professions. As our collective understanding improves, we may also leverage new training regimes. But most of all, we should be building better worlds in which to train the next generations of predictive minds.

We *can* learn to "expect better." And we must.

6

BEYOND THE NAKED BRAIN

TABITHA GOLDSTAUB is a successful tech entrepreneur. She is also a dyslexic who, as a child, had a great fear of numbers and words. Nowadays, that fear is gone, replaced by a joyful optimism and an enviable writing style. In a recent newspaper article, she describes her situation like this:

> I rely on apps such as SwiftKey and Grammarly as one might an old friend. SwiftKey in particular is a huge help in my day-to-day life. It's an app for your smartphone keyboard that uses AI to make much better recommendations than the inbuilt spelling and grammar check. Even better is its new feature that turns my voice to text so I don't have to type or leave a voice note when I'm struggling to find exactly the right way to say something. Grammarly is my go-to for my laptop. It combines rules, patterns, and AI deep learning techniques to help you improve your writing. [But] if something goes wrong with either of these apps, I feel as I'm back in the classroom again, freefalling, my brain foggy, letters and numbers jumbled up.

Goldstaub is a prime example of what is sometimes called an "extended mind"—her normal, daily mind is not the one realized solely by the dyslexic hardware. Instead, that inner hardware is now robustly coupled, for most of the time at least, to various external technologies. I believe that it is the resulting coupled systems combining brain, body, and technology that we should then recognize as the true incarnation of Tabitha Goldstaub's mind.

But this isn't really about Goldstaub. It's about everyone. We all rely, to greater or lesser degrees, upon a wide array of apps, tools, and other "beyond the brain" resources to carry out our daily projects, to organize our lives, and to remember to do things that we'd otherwise forget. Some of these aids function in ways that seek to replicate or bolster skills and abilities already possessed by our biological brains. Others play even more intriguing roles—not simply replicating native biological capabilities but enhancing and transforming them.

When the coupling with key tools and technologies is robust and reliable, so that the brain learns to simply expect the presence of those resources, factoring their effects into all our planning and actions, we become (I'll argue) extended minds—cyborg or hybrid minds created without the need for invasive implants. The machinery of our minds, despite the lack of any Terminator-style circuit boards inside the skull, is then no longer exhausted by the operations of our biological brain alone.

In this chapter, we'll discuss this provocative theme from two neatly interlocking perspectives. The first perspective highlights the importance of "epistemic actions"—actions selected to improve our state of knowledge rather than to directly achieve some practical goals. If catching the plane is your practical goal, a useful epistemic action might be to check bus times before setting off. The second perspective is thinking about action—including epistemic actions—as controlled by prediction. Bringing these perspectives together reveals predictive brains as ideally poised to discover epistemic actions, allowing them to make good use of information-bearing resources (such as bus timetables) in the wider world. I'll try to show that predictive brains do not care whether key information is stored in their own internal states and structures or outside, in notebooks, apps, and GPS systems. What matters is just that the right information or operations are predictably available as and when needed for the fluid control of behavior. The result is a delicate dance in which inner and outer resources constantly cooperate: a dance in which prediction-hungry brains provide the perfect biological platform for extended minds.

Leaning on the World

My first full-time academic post was in the mid-1980s, as a member of the Cognitive Studies Program (as it was then called) at the University of Sussex in the U.K. This was a pioneering unit in the field now known as Cognitive Science. The Cognitive Studies Program had been founded in the early 1970s, and Sussex lays claim to being the first university in the world to have degrees in cognitive science on its books. I was amazed and delighted, amidst the turmoil of Margaret Thatcher's relentless attacks on the U.K. education sector, to have somehow secured a real job. Not only that—it was in my ideal interdisciplinary program, in the vibrant seaside town of Brighton, on the relatively sunny south coast of England. I was over the moon.

Things got even better. Very soon after my arrival at Sussex two volumes appeared that had a major impact on my life and career. They were the two volumes of the "connectionist bible" entitled *Parallel Distributed Processing: Explorations in the Microstructure of Cognition*. Published in 1986 by MIT Press, they were my first sustained introduction to work that would nowadays be called simply "artificial neural networks." I can see those two hardback volumes (one blue, one brown) right now on my shelf. They were a very expensive purchase for a young academic. But they were worth every penny.

One paper in the volumes stood out for me. The puzzle that the paper sought to tackle was just how our brains could manage to solve whole classes of puzzles that taxed—and perhaps even exceeded—their native capacities. The solution was as obvious as it was neglected in the literature. At its core was the simple observation that we humans would often tackle such puzzles by recruiting external props and tools, in many cases reducing the more complex puzzles to sequences of simpler ones that our biological brains could more readily handle.

For example, confronted by a long multiplication (such as 77777 times 99999) many of us, myself included, would in those days resort to pen and paper. The reliable availability of such resources in our daily lives means that we can usually manage to solve complex problems simply by training our brains to solve much simpler ones, such as 7 × 9, and by creating a certain procedure for writing down the results in a way that allows repeated iterations of these simpler calculations to solve the puzzle. Here, culturally transmitted practices change the problem space, so we can do more with less, easily solving an open-ended set of such multiplications many of which would defeat even the most skilled naked human brain. The paper included, as a kind of throwaway aside, the observation that "On this view, the external environment becomes a key extension to our mind." It's safe to say that I took this suggestion to heart.

Much of my own subsequent work on the embodied and extended mind has been a sustained exploration of this simple but compelling idea. One implication is that what goes on inside the head might often be simpler than we have imagined. It might also be simply different, in that much of the real work of the embodied brain now consists in learning the right strategies for interacting with the external world. It is here that the predictive brain excels. As it does so, these problem-solving loops (in which the brain leans upon external props and resources) become more and more part of our daily routines. Our minds, actions, and worlds meanwhile become more and more closely entwined. Understanding this process reveals the human mind as a "leaky system"—a system apt to lean on the surrounding world in heavy and sometimes unexpected ways.

Over cultural-evolutionary time we have built a world of artifacts and tools that very neatly complement the capacities of our biological brains, allowing us to lean on the world in ever-more-complex ways. But even the simple pencil provides a stable, robust, real-time responsive means of offloading key intermediate results of processing onto a notepad, a loop we can repeat again and again as we build up more and more complex thoughts and ideas. As that process unfolds, brain, body, pencil, and notepad act as a new and potent whole.

We are so familiar with this kind of entwining that we mostly cease to notice it. An architect using brain, sketchpad, and a suite of powerful apps is a design powerhouse. When seeking a good design, we seek out a good architect. But we would never ask that the architect show us their true prowess by removing their apps, pencils, and sketchpads!

Home Alone

The potential depth and importance of cognitive entwinement was brought home to me when, in late 1993, I first moved from the U.K. to the United States to direct the Philosophy-Neuroscience-Psychology (PNP) Program at Washington University in St. Louis. My upstairs neighbor, in a lovely old brownstone off the green spaces of Forest Park, was Professor Carolyn Baum, then head of occupational therapy at the Washington University School of Medicine. In conversation, Carolyn mentioned something that had been puzzling her when testing people diagnosed with Alzheimer's, who were living in central St. Louis. Despite the severity of many of these cases, they were frequently found to be doing well despite living alone in the inner city. Yet their performance on all the standard tests and protocols suggested they should simply not have been able to cope.

When Baum visited them at their home addresses, the riddle was solved. Their homes were a model of what would today be called "supportive environments." These were homes full of helpful props and aids that had been designed either by the dementia sufferers or, quite often, by their loved ones. These included message centers, where they stored notes about what to do and when; numerous photos of family and friends with notes of names and relationships; internal doors and cupboards that were complete with labels and pictures; memory books, in which to record new events, meetings, and plans; and open-storage strategies—meaning that crucial items such as pots, pans, and checkbooks (it was 1993 after all) were always kept in plain view. They were also using simple daily routines such as getting the same bus to and from the same store to buy essentials and food.

The overall effect was that their homes and local environments took over some of the functions that were previously played by their biological brains. Thanks to these redistributions of cognitive labor, these folks were able to live successfully in the challenging inner city. Nowadays, there is widespread recognition of the role of "dementia-friendly environments," with many websites and care homes actively seeking to improve the lives of dementia patients in these ways.

Now try the following thought experiment. Imagine a world in which biologically normal human brains all function like the brains of those with these forms of dementia. In that alternative reality, we would have surely evolved technological and social structures, like those mentioned above, that would work successfully given those biological endowments. Perhaps we would have slowly evolved (by historical trial and error, and thanks to a few potent innovations along the way) a form of life in which the various kinds of prop, scaffolding, and social practice that supported Baum's inner-city dwellers was simply the norm. With that thought experiment in mind, take another look at the stuff that currently surrounds you in your own life and work. There may be notebooks, smartphone apps, GPS devices, and more. All these elements come together to enable you successfully to pursue your own life and projects. This shows that normal healthy brains lean heavily on a wide variety of nonneural props and resources too.

There is a real sense, when we appreciate the extent of this reliance, in which we humans are, and long have been, what I once called "natural-born cyborgs." A cyborg is an entity that is part human, part machine (or technology). But it's a mistake to think that human-machine hybridization must always depend on wires and implants. Instead, what matters is the reliance of the biological parts on capacities made available by the nonbiological parts. Seen from that angle we are indeed cyborgs—beings whose daily abilities to plan, reason, and decide are already realized by complex webs of biological and nonbiological structures. Of course, our brains must still support the many skills needed successfully to engage those "outer" resources. But the suite of capacities that make us who and what we are is often best understood—or so I argued—as the larger hybrid whole.

In the case of the inner-city Alzheimer's patients, their home environments had become effective scaffoldings that worked to offset some of their inner cognitive compromise. So effective are these scaffoldings that when they are removed (as often happens when, for example, the person is suddenly relocated to a care home) mental health and the ability to cope can plummet. The damage can be somewhat repaired by modeling the new environment on the old, for example by painting the door to their care home room in the distinctive colors and patterns of the home front door. But such measures inevitably fall short, and there is a strong sense that something quite profound has often been lost in the transition—that they have been deeply mentally and emotionally compromised, even when there really is no other alternative. This is gut-wrenching evidence of the surprising fuzziness of the self/world boundary. In a certain sense, our minds and selves can become so intimately bound up with the worlds we live in that damage to our environments can sometimes amount to damage to our minds and selves.

In fact, we have probably all experienced mild versions of this when we lose or misplace our smartphone. The suite of abilities upon which we normally rely on is suddenly diminished. We feel a bit lost as we attempt to confront our daily tasks without our timers, reminders, and GPS—just like Tabitha Goldstaub when her apps were down.

Acting for Information's Sake

Predictive brains are the biological engines that make all this deep entwining possible. To see how, it helps to start by returning to the distinction between practical and epistemic actions.

Our lives are filled with actions, but we often think of those actions mostly in terms of their practical goals. Perhaps I am trying to win at Scrabble, or to make and serve cocktails in a busy bar. But if we look a little closer, it becomes clear that (just as in the case of long multiplication) many of my actions serve these goals in an interestingly indirect way. They serve them by improving the information stream available to the brain, or by altering the problem space itself. We'll see several examples as the chapter progresses.

A very simple one is this. The skilled Scrabble player

physically shuffles the tiles, not because this is itself a way to score points (if it was, I'd be a much better player than I am). Instead, they shuffle the tiles to prompt the biological brain with new potential word fragments—fragments that might better prompt the neural system to recall a high-scoring word. Reshuffling XEO to EXO can prompt you to consider EXORCISE as a candidate word, and to check your hand and the board for ways to leverage that whopping 76 points.

Or consider the expert bartender. Faced with multiple drink orders in a noisy and crowded environment, they mix and dispense drinks with amazing skill and accuracy. But what is the basis of this performance? Does it all stem from finely tuned memory for drinks orders? Not exactly. In controlled psychological experiments comparing novice and expert bartenders, it was found that expert skill involves a delicate interplay between brain-based and environmental factors. The experts select and lay out distinctively shaped glasses at the time of ordering. They then use these as environmental cues to help recall and sequence the specific orders. As a result, expert performance plummeted in tests involving uniform glassware, whereas novice performances, though much worse overall, were unaffected by the change.

The expert bartender has learned to sculpt and exploit their working environment in ways that transform and simplify the task that confronts them. The problem of remembering which drink to prepare and serve next was transformed into the much easier problem of perceiving the different shapes of the cocktail glasses (each being associated with its own drink) and noticing their place in the line. In this way—as we already saw in some more basic cases back in Chapter 3—the exploitation of environmental structure allows relatively lightweight cognitive strategies to guide complex and thoughtful-seeming behaviors.

In classic research from the late 1990s, the cognitive scientists David Kirsh and Paul Maglio showed in wonderful detail how expert players of the computer game Tetris combined practical and epistemic actions fluently, sometimes rotating a descending geometric shape (a zoid) to aid identification rather than to fit it into a waiting slot. The seamless intermingling of these epistemic and practical moves in expert play suggested the operation of a single overarching strategy—which is exactly what work on the predictive brain now provides.

Epistemic (knowledge-improving) actions are chosen not because they are of intrinsic value to us, nor even because they move us closer, physically speaking, to some practical goal. Instead, they may even move us temporarily further away. For example, if I'm driving, I might navigate back to a familiar spot that I know is in entirely the wrong direction, if I happen to know a reliable route from that spot to my destination. This is sometimes called the "coastal navigation algorithm," since a sailor may navigate to the coast in order to better find their way, even if following the coast is a much longer route. This renders the distinction between epistemic and practical components especially sharp. But in many other cases (such as the selection of a certain glass during the cocktail-making procedure) the distinction between the epistemic and non-epistemic elements becomes fuzzy, and potentially vanishes entirely. We'll now see how this fuzziness makes sense from the perspective of the predictive brain, which is simply trying to minimize error in the pursuit of long-term goals.

Unifying Practical and Epistemic Action

Consider Mego, the orangutan shown in Fig. 6.1. Mego has learned to probe the waters with a stick to determine depth before attempting a river crossing. Orangutans are famously adept tool users, so much so that Michelle Desilets (executive director of the Orangutan Land Trust) likes to quote the saying that "if you give a chimpanzee a screwdriver, he'll break it; if you give a gorilla a screwdriver, he'll toss it over his shoulder; but if you give an orangutan a screwdriver, he'll open up his cage and walk away."

The use of the screwdriver is a practical action, while the use of the stick to determine depth is an epistemic one. But neither the orangutan nor the orangutan's brain needs to mark

Fig. 6.1 Mego the orangutan performing an action designed to improve his state of information

that difference. Instead, both strategies emerge directly from the attempt to minimize a quantity known as expected future prediction error.

That's a somewhat clumsy phrase I fear, though it is the one popular in this literature. It is clumsy because the whole idea of expecting error can seem puzzling. But it just means having a capacity to look toward the future and compare what will happen if we were to take one versus another course of action. You then choose the actions that best resolve the errors (the "expected future errors") that would otherwise arise. For the most part, we are not even aware this is happening. When you google movie times, you are probably not aware that you are resolving ambiguities about the future, acting to minimize errors that would otherwise occur. But you are doing just that. In the case of Mego, the best way to reduce key uncertainties (and so avoid future error) was to probe the water with a long stick before taking the plunge.

In all these cases, practical actions and epistemic actions are determined in exactly the same way, as the predictive brain makes counterfactual predictions about what kinds of futures will result if certain actions are launched. Actions are then chosen that deliver preferred outcomes directly (when possible) or else that probe and sample the environment to bring forth more information, reducing key uncertainties, and making the desired outcome more likely in the future. In other words, practical and epistemic actions both flow from that same deep source and serve a common goal. The difference between the orangutan and the human lies mostly in the depth and general nature of the world-knowledge they bring to bear—a human might, for example, create a time and tide table that enables her to plan river crossings for many months ahead.

Any predictive processing agent able to minimize error relative to future goals will discover both epistemic and practical actions, and how to mix them together. In all such scenarios, all the brain does is select the actions that best minimize future prediction error relative to goals—where the goals themselves, as we saw in Chapter 3, are really just strong (precise) predictions that describe desired future states. Once a system can compute expected future error, it will automatically seek out the interwoven set of practical and epistemic actions most likely to bring the desired future state about. In this way, predictive brains repeatedly spin extended problem-solving webs that combine practical and epistemic actions, pulling in key resources such as pencil, paper, apps, smartphones, and notebooks at the right moments.

The ability of prediction error minimizing systems to find solutions of this kind has now been demonstrated in multiple studies, including one in which simulated rats discovered the mixture of practical and epistemic actions that would best enable them to find rewards in a maze. The simulated rats started each run at the center point of a simulated three-arm (T) maze in which food rewards (preferred states) could be found at the end of one of the two arms—the right and left parts of the top of the T. The lower part contained a cue that tells them where to find the reward on each trial. Rather than directly explore each upper arm, the rats learned to navigate temporarily away from their targets to the lower location that never actually contained a reward but that always contained useful information. Here, just as in the "coastal algorithm," the simulated rats moved away from the known possible food locations, navigating instead to the place from which (thanks to the knowledge-improving cue) a reliable route straight to the reward was assured.

By first going to the cue location, the rats gain information enabling them to plot efficient ways to the food sources that (in the real world) would help maintain the body budget essential to life. This reminds us that trade-offs between epistemic and practical action emerged early in the history of life, even though they have become more striking as our knowledge has increased in scope and depth, and as helpful technologies have taken root and multiplied. Now, we might use Google to help find a restaurant, one serving up rewards (such as unusual sashimi dishes) that are at best tenuously linked to metabolic necessities. But the dance between practical and epistemic action remains the same.

Looping Processes

Imagine you want to build a new kitchen for your home. To do so, predictive processing suggests, you start with the expectation (the "optimistic prediction") of a good outcome. Your task (or one of them) is then to select epistemic actions that resolve key uncertainties such as where to put the cooker, dishwasher, and fridge. This is often best achieved by combining mental actions such as imagination with physical ones such as looking at catalogs, measuring, sketching, and remeasuring. In contemporary settings you might also use an app to try out various items from different kitchen ranges. The app provides a kind of augmented imagination that improves considerably upon the biological "mind's eye," enabling us to spot opportunities and problems that we would otherwise miss.

Designing a kitchen involves a complex mix of epistemic and practical actions. In creating that mix, timing and sequence matter too. This is perhaps best appreciated by thinking about another case—that of fast-paced sports. By having a good predictive model, a player can preemptively look toward locations where there is currently nothing going on but where they predict that crucial information is about to appear. In soccer, for example, a player will often look to the spot where they think a pass will soon be made. The control of these epistemically motivated head and eye actions must depend on predictions rather than here-and-now perceptions, as there's often absolutely nothing of any interest going on at that location just yet. These information-seeking forays must often be launched at just the right moment if they are to lead to sporting success.

This kind of predictive skill develops with expertise. Learner drivers, for example, don't predict road events as far ahead as experts. But they are actually better at spotting new events that are highly unpredicted, as their visual scan paths are less bound by strong ingrained expectations about where and when important things are most likely to occur. The expert driver, by contrast, has learned how to look to just the right place at just the right time, in ways that work extremely well, far outperforming the novices most of the time. But that same expert may fail quite spectacularly when truly unusual events occur, such as a cyclist suddenly entering a roundabout from the wrong direction. In such cases, their expert prediction systems cause them to fail to scan the whole scene.

Timing matters while using cognitive aids such as pen, paper, and app too. When I think while writing things down, count using my fingers, or create a design using an app, all the elements (neural activity, bodily action, and responses in the external media) entwine, each seeming to inform the other at just the right moment. But add a few unexpected time delays (as sometimes occurs when, for example, your broadband connection is bad) and your ability to "think via the keyboard" rapidly dissolves. This is because the brain predicts certain speeds and latencies when using specific tools and technologies, and those predictions launch epistemic actions at just the right times to serve our needs. This is what weaves inner and outer operations into fluent, extended problem-solving wholes. When all goes well, this results in such a seamless, loopy integration that we start to feel that we are really thinking via the extended routine (scribbling, sketching, or using an app) itself.

This feeling of seamless integration is beautifully captured by the Nobel Prize–winning theoretical physicist Richard Feynman in a famous exchange with the historian Charles Weiner. Weiner had suggested that a certain batch of notes and sketches were a useful record of Feynman's day-to-day work. But Feynman reacted characteristically sharply:

"I actually did the work on the paper," he said. "Well," Weiner said, "the work was done in your head, but the record of it is still here."

To which Feynman replied "No, it's not a *record*, not really. It's *working*. You have to work on paper and this is the paper. Okay?"

Feynman here instinctively recognizes that his mathematical thinking emerges not simply from the activity of his brain but from the whole embodied cycle—a cycle in which scribbling things down as you go along plays a core role. That writing down is not simply leaving a record, but part of the actual process of thinking things through. The process of writing this book can be seen in this way too. It is not as if the text somehow springs fully formed from my brain. Instead, my brain acts as a constant facilitator of a stack of repeated interactions with various external resources. As these resources (old notes, key articles, web pages, and online discussions) are encountered, my brain reacts in a fragmentary way to each, very occasionally delivering new ideas that lead to further notes and scribbles. These are repeatedly refined, re-encountered, and transformed in what is best seen as a rolling, extended process of thinking and text construction. In these ways, many of our prime cognitive achievements should not be credited solely to our biological brains but depend heavily upon the enabling environments in which we act and perceive.

A Chip off the Mainframe

Despite this, not every extended problem-solving process should count as a case of extending the core machinery of an individual mind. Sometimes, a tool is just a tool, an app is just an app. In many of the cases we have considered so far, the distinction between the thinking (cognitive) agent and the supportive environment remains clear. But there are parts of the world that are with us so constantly, and that function so reliably, that the biological brain can come to treat the capacities they provide as a kind of given. In such cases the brain automatically factors in our technologically augmented capacities in much the same way it learns (as we saw in Chapter 3) to take our basic bodily capacities for granted.

To take a homely example, we may learn, as children, that our fingers are always there and can be used in the service of keeping track of numbers. Our brains may then develop counting strategies that simply rely on the presence of the fingers and are incomplete as problem-solving recipes without running action loops through the fingers. Some of our mobile devices are now positioned in rather a similar way. They are usually available, and from a very young age our brains learn to factor in the suite of capabilities they pretty much constantly provide.

Imagine next that part of your brain is malfunctioning and a replacement for that part gets created using a silicon chip implanted inside your head. If the chip exchanged signals with the rest of the brain in just the right way, most of us would accept that the chip has now taken over the missing functionality and is working just like the damaged part of your brain. As such, that chip is now part of the material underpinnings of your mind, enabling you to once again solve the kind of problems for which the damaged functionality was key.

Now vary that thought experiment just a little, so that the chip is stored externally but is in constant wireless contact with the rest of your brain. Your abilities would be similarly restored, so surely the functionality of the externally located chip should still now count as part of the machinery of your mind? To go that far, however, is to open the door to the radical vision of minds whose machinery exceeds that of the individual brain even when some of that additional machinery is not attached to, or implanted within, the body.

We can creep up on this with an analogy—one that starts from within the body but will help us make sense of these more radical consequences.

Thinking from the Gut

Consider coalitions of neurons that are already located outside the brain. An increasingly familiar example can be found inside the human gut, where upward of 500 million neurons in the gut wall already relay important information to the spinal cord and the brain. This circuitry helps regulate serotonin and other neuromodulators. The so-called gut-brain is by a long margin the largest cluster of neurons outside the brain, and an essential part of the nervous system. It is pretty clearly part of what makes you who you are and has a major influence on what you think and feel. This already gives the lie to the idea that your mind consists entirely of "what the brain does."

But there's more. Our gut is also alive with (mostly) helpful bacteria, which together comprise the "microbiome." These gut bacteria (unlike the neurons) are not even "genetically you." But they too make essential contributions, and have been shown to affect learning, memory, and mood as well as basic bodily regulation. Such links are not surprising given the deep role (recall Chapter 4) of bodily information in the construction of the mind. For example, gut bacteria manufacture up to 95 percent of the body's serotonin, which has large impacts on mood and is one of the neurotransmitters implicated in the precision-weighting process.

In one striking experiment, mice that were specially bred to be unusually timid were given doses of an antibiotic that radically altered the composition of the bacterial colonies in their guts. The mice with the altered bacteria then became bold and risk-taking. There were also increases in levels of a neurochemical that helps learning. Once the antibiotic treatment ceased, the mice reverted to their usual timid selves. Cementing the picture, the same team performed a similar experiment using two strains of mice, one of which was bred to be timid while the other was aggressive. Under the influence of the antibiotics, these roles were reversed. This showed that what looked like genetically determined behavioral profiles in the mice was actually deeply dependent on the gut bacteria too.

The complexity of these links was further evidenced in experiments involving infant monkeys. In these experiments, monkeys whose mothers had been startled by loud noises during pregnancy had reduced levels of specific kinds of gut bacteria associated with calm, anxiety-free moods. In the monkeys, stress-induced changes to the composition of the gut microbiome (passed from mother to baby by entirely nongenetic means) were part of a mutual feedback process, in which the mother's experienced stress altered the microbiome in ways that would lead to even greater anxiety in the baby. Successful organisms are thus much more than instances of simple genetic lineages. They are, as the case of the microbiome neatly shows, collaborative ecologies that require a wide variety of contributions.

Life, as the philosopher of science John Dupré and the philosopher of microbiology Maureen O'Malley once put it, is always a deeply collaborative affair.

Extended Sensing

Let's return to that imaginary wireless chip that restores brain function and change what it does so that it is not simply an off-brain version of a damaged neural circuit but contributes—rather like the microbiome—brand-new kinds of functionality. Perhaps it constantly tracks part of the stock market and sends you an alert whenever key indicators suggest increasing volatility or (because it is also monitoring the newsfeeds) if major political events or natural disasters are being forecast. The alert could take the form of a sudden mild shock or tingle. Once alerted, and if more detail is needed, that could then be supplied—for example via some kind of complex visual overlay in augmented reality.

We could increase the intelligence of the chip too. Suppose now that the device learns to vary both its sensitivity and the key markers that it is responding to according to recent market trends. It could even factor in (via some kind of wearable interface) your own bodily background states, thus avoiding nonessential communications if you appear stressed or tired. Over time, your brain learns to rely heavily on signals from the chip to indicate actionable, exciting trading opportunities.

You now have a new brain-body circuit—a kind of spider-sense, for us comic book fans. Just as night vision goggles enable us to see in the dark, so these "stock market goggles" enable you to sense market volatility and opportunities. Over time, the functionality of the chip becomes (or so I am arguing) as much part of you as the functionality of all those neurons in your gut. So good is the merger, in fact, that your brain now relies on signals from the chip as fluidly and automatically as it relies on the deliverances of your ordinary senses. The chip has become what I will now call a "woven resource"— a trusted nonbiological structure that is delicately and constantly coupled to the rest of the system that we recognize as "you."

Early versions of such deep-weave technologies are with us already. A simple example is the North Sense—a small silicon device that is attached to the chest and that delivers a short vibration when the user is turned toward magnetic north. This constant drip feed of directional information is rapidly assimilated into the cognitive ecology of the wearer, who soon simply expects to know, moment by moment, their orientation relative to important distant places such as their home or their children's school gates. In this way, the North Sense couples with our affective response circuitry too. All that functionality is now simply taken for granted by the predictive brain, and users report varying degrees of anxiety and distress when the device is later removed or becomes inoperative.

Many of our daily devices, especially our smartphones and other wearables, are already starting to act as woven resources. They are devices whose constant functionality has become factored deep into our brain's ongoing estimations of what kinds of operation can be performed and when. When they are also capable of monitoring the physiological states of the user, this creates potent new brain-body-world circuits. Near-future technologies will surely take this to a whole new level.

In my previous work, and in my more recent role as an occasional academic consultant with Google UK, I found myself imagining a future in which human beings are slowly surrounded, from quite a young age, by layer upon layer of intelligent personal and household devices. Progress here will be accelerated by the proliferation of "edge computing" in which information storage and transformation takes place closer to the origins of the data. This will enable many problems to be solved "on device" (without sending data all the way up to a distant cloud server and back) using data generated by sensors and users as they are moving about the world in real time.

In these future worlds, we humans develop from the outset

in the presence of a suite of supportive technologies. Among these will be personal AIs—some living entirely on-device, some perhaps woven into our clothing, some others (not quite so personal) that permeate the wider environment of roads, homes, transport, and offices. Your personalized AIs would come online when you are very young. They would learn from your choices and contribute to your choices in turn. They may also help with meta-tasks, highlighting and recruiting other resources—ones currently living just outside your personal ecosystem—to help you approach your goals. Living, working, and playing in these enriched settings we will continue the humanity-defining process of blurring the already fuzzy boundaries between self and nonself, mind and tool, person and world.*

Extended Minds

The philosopher Jerry Fodor once wrote, "If the mind happens in space at all, it happens somewhere north of the neck." Fodor emphatically rejected the idea that the machinery of individual human minds could include goings-on in the rest of the body (the bits south of the head) or, worse still, the wider world.

The extremely heretical view to the contrary was pioneered by myself and David Chalmers in a short paper written in the early 1990s, back when I was directing the PNP program at Washington University in St. Louis. One of my first acts as director had been to persuade Chalmers to join us as a postdoctoral fellow—a research-heavy role ideal for such an academic rising star. Dave is now famous for his work

* These brief reflections upon imaginary futures and increasing degrees of human-machine symbiosis reflect my own thoughts and speculations and not Google plans or policies.

on consciousness and (most recently) on how we should think about virtual and augmented reality. But our short paper has become a kind of modern classic and remains one of the most cited papers in contemporary philosophy of mind.

The paper was called "The Extended Mind" and in it we argued that the machinery of an individual mind did not have to be restricted to the machinery of that individual's brain and central nervous system. It did not even have to be restricted to their body more generally construed. Instead, true mental circuitry could indeed be spread out across brain, body, and aspects of the material and technological world. The idea was that under certain conditions outward loops that involve quite mundane goings-on (such as consulting calculators or smartphones or even just looking at things we've written down in notebooks) could count as proper parts of the machinery of thinking. Your mind, we argued, isn't always all in your head.

At the heart of our argument lay a very simple theme—one that already looms large in our discussions. It is that one of the functions of the biological brain is to create and maintain perception-action loops that keep us alive and that bring us closer to our goals. This requires both storing information using "onboard" memory, and also actively seeking out additional information as and when it is needed. In this search for good information, it doesn't matter whether the information is already stored in memory or requires the use of bodily actions that loop in various tools and technologies. What matters is just that the right information becomes available at the right moment. Our radical suggestion was that when the weave between the brain's activities and the functionality of some nonbiological resource becomes sufficiently tight, it really is better to think of that person's mind as an extended mind—a new problem-solving architecture built from an array of resources spanning brain, body, and world.

The argument we presented involved a general principle

(the parity principle) which is best seen as a heuristic, a rough-and-ready tool, for identifying plausible cases of cognitive extension. The parity principle went like this:

> If, as we confront some task, a part of the world functions as a process which, were it to go on in the head, we would have no hesitation in accepting as part of the cognitive process, then that part of the world is (for that time) part of the cognitive process.

The idea here was simply to invite the reader to judge various potential cognitive extensions without the distractions of location (is it in the head?) and human biology (is it made of wet stuff?). A good way to do this is to ask yourself, concerning some potential cognitive extension, whether if you were to find that functionality operating *inside* the head of some alien organism, you would tend to count it as part of the machinery of the alien mind? If so, then the onus—we claimed—is on the skeptic to tell us why that same process, when making its contribution from outside the brain, should not count as forming part of an extended mind.

Apply that reasoning to the responsive stock market chip discussed earlier, and you will probably conclude that that functionality, if it were found to have somehow developed inside an alien brain, would unquestionably count as part and parcel of the machinery of that alien mind. To then think differently about the wireless-communicating version (the externally located chip) would be just some kind of unprincipled "neuro-chauvinist" prejudice. The same seems true of a B-movie scientist's trusty calculator or slide rule, although these communicate with the brain by means of whole action-perception loops that involve the usual sensory channels. We do not think this difference makes a difference. In each case, fluent integrated functionality is achieved. Perception-action loops can thus enable normal human minds to become extended minds—minds that include some nonbiological parts.

If this is right, then Fodor was wrong. Minds are not merely what brains do. They are what brains *create*—distributed cognitive engines spanning brain, body, and world.

Otto Goes to MoMA

In the original paper, we used the example of a mildly memory-impaired agent (Otto). As his memory declines, Otto has increasingly relied on a notebook that he always carries with him. In the notebook, he writes down addresses, facts about his family, friends, the wider environment, important dates, etc., so that the notebook has come to play a role that's similar to his memory. One day, Otto decides he'd like to visit MoMA—the Museum of Modern Art. To get to the museum, Otto consults his trusty notebook, and reads that it is on 53rd Street.

Now imagine someone named Inga, whose onboard memory is working well. Inga also wants to visit MoMA. But she simply recalls that MoMA is on 53rd Street. We would normally say that Inga already knew, even *before* she retrieved it from her memory, where the museum was situated. But what about Otto? The retrieved fact "MoMA is on 53rd Street" was not stored in his brain, but in his notebook. But that information was easily accessed, at just the right moment, to guide his behavior. So—in line with the parity principle—we argued that even *prior* to looking at his notebook, Otto too should count as knowing that MoMA is on 53rd Street. Otto's notebook was functioning as part of his extended memory and playing something quite akin to the role of Inga's memory.

Another way to look at this (drawing upon the earlier discussion of epistemic actions and predictive brains) is to consider Otto and Inga as both engaging in a kind of foraging-for-information. Inga's foraging explores only her own onboard memory, while Otto's foraging explores external information sources such as the notebook. But these have something deep in common—they are both ways to reduce expected future prediction error so as to bring us closer to our goals.

Some have argued that the involvement of perception-action loops in cases such as Otto's use of the notebook or our use of a smartphone marks a crucial difference. Otto has to perceive and act upon the notebook to discover the necessary information, whereas Inga simply retrieves it without "looping out" through perception and action. Inga's epistemic actions (retrieval from memory) are purely internal ones, whereas Otto's are not. So perhaps it is perception and action, not skin and skull, that best serve as the boundary of the mind? We (myself and Chalmers) are not convinced by this move. Our current view is that the true core of the extended mind thesis lies precisely there, in the claim that the boundaries of perception and action need not mark hard-and-fast boundaries of the "mind."

When we wrote "The Extended Mind" in the 1990s, mobile computing technology was not nearly as developed as it is today. Today, the best illustration of our thesis would probably involve a smartphone. As Chalmers notes, hardly anyone relies on their memory to remember phone numbers anymore. The smartphone is playing the same role that our memory used to play. According to the extended mind thesis, our smartphones have—in those respects at least—already become part of our minds.

It is intriguing to note that a whole class of artificial neural networks systems (called differentiable neural computers, or DNCs) are now emerging that rely on a form of "extended memory" too. DNCs are artificial neural networks that couple their own internal processing capacities to stable yet modifiable external data stores. These "extended computing" systems can reason about a variety of complex problem spaces—such as how to navigate the London Underground network—by coupling their processing to various kinds of external information stores, such as a London tube map. These systems exemplify, in a minimal but revealing fashion, the way that information foraging loops into the world of stable, rich external storage can function as parts of extended computational processes.

A Snapshot of the Debate

You might worry that there are important differences between the way memory works in Inga and Otto. But there are also differences in the way memory works in different species. Memory in foraging honeybees, for example, seems to be quite different from memory in humans. The underlying architecture of the bee brain is not the same as that of mammals, but that does not mean that honeybees cannot remember. They can recall the location of nectar and find efficient routes back to their own hive. Despite having only a very tiny brain, they can use memory to plan their activities, and they can adapt flexibly to new information as it arrives.

Another common response to our little thought experiment is to say that Otto's notebook cannot be part of his mind because it is outside his head, or perhaps because it is not biological. But this seems nothing more than a sort of "skin and skull" biological chauvinism. Skin and skull do not, as one of my all-time favorite philosophers, Susan Hurley, once neatly put it, form some kind of "magical membrane" uniquely suited to act as a privileged boundary for the machinery of mind.

Yet another common objection is that all Otto knows in advance of the moment of retrieval is not the actual address of MoMA but only that the right information is stored in his notebook. By contrast Inga, even before accessing memory, already counts as having the full belief that the museum is on 53rd Street. Here too, we argue for parity of treatment. If you insist that all that Otto believed prior to accessing the notebook was that the address was stored in the notebook, then you should also say that all that Inga believed before retrieving the address was that the information was stored in her biological memory. We don't usually say this about Inga, of course. But nor, we think, should we say it of Otto. Otto's notebook-consulting behavior is now part of a habit system that is usually engaged without his needing to consciously think to himself, "I will now access my notebook to look for the address." Otto reaches for the notebook as thoughtlessly and automatically as Inga "reaches" for her biological recall.

Maybe it seems to you that the notebook cannot form part of the machinery of Otto's mind since it is not sufficiently "essential" to Otto. He might lose the notebook, but he'd still be Otto. But this kind of reasoning is also a mistake. My visual abilities are currently part of me, part of what makes me the cognitive being that I am. But I could survive their loss and still be me. In any case, onboard memory is fragile too, as the case of Otto himself suggests! Inga, after one martini too many at lunchtime, might be temporarily unable to remember MoMA's address. Once again, rough parity between the cases seems to rule the day.

Likewise in the case of Tabitha Goldstaub. By isolating her from the Grammarly and SwiftKey apps, you can impair her performance. But you can also impair my brain-based performances by giving me a soporific drug, or (as some researchers have done) by applying a magnetic pulse to my brain. My capacities might also change following a stroke. In none of these cases do we think that the degraded performance determines what should have been counted as my "true mental machinery" all along.

As the range and use of assistive technologies for those with various biocognitive impairments increases, it will become more and more important to recognize this point. This also means that deliberate damage to closely woven assistive technologies should be regarded much like deliberate damage to your brain. Recall Goldstaub's comment that when the apps are down, she feels foggy, free-falling.

Solving the Recruitment Puzzle

The original notion of the extended mind was developed without the benefit of a solid account of what brains do. As a result, it left dangling the question of how the right mix of actions, internal and external, come together at the right time. Somehow, the canny biological brain manages to recruit, activate, or exploit, on the spot, whatever mix of problem-solving resources will yield an acceptable result with a minimum of effort. But how?

The broad shape of an answer to the recruitment puzzle is now clear. It is predictive brains, I believe, that explain and make possible the looping encounters that build extended minds. We have seen that predictive brains constantly estimate the extent to which taking an action (such as using the stick to probe the depth of the river) will reliably reduce uncertainty in ways that help us approach our goals. The ability to make these estimations reflects the operation of a counterfactually rich predictive model—one that looks ahead to see what we should expect to experience if we were to perform different actions. This enables the selection of whole sequences of actions that allow us to steadily approach our goals.

Epistemic actions emerge naturally as part of this process. This is because actions in the here and now can be selected to gather information that improves our chances of future success. In the case of Otto, that means selecting the action of consulting the notebook (to get to MoMA, which is where

Otto's brain "optimistically" predicts he will shortly be). In the case of Inga, by contrast, it means internally retrieving the address from onboard memory. But each of these moves now arises in the same way and for exactly the same reasons. They are epistemic actions (internal or external) that arise because the predictive brain is seeking to minimize future prediction error.

The predictive brain is here engaging in a kind of "knowledge budgeting" akin to the "body-budgeting" activities that we met back in Chapter 4. Knowledge budgeting involves selecting policies and actions that will steadily, and at just the right moments, deliver the knowledge and information needed to approach our long-term goals. That, roughly speaking, is how predictive brains solve the recruitment puzzle. It is solved by estimating which actions and policies best resolve key uncertainties and hence reduce the distance between our "optimistic predictions"—such as arriving safely at MoMA—and our current state. Likewise, the person who predicted arriving at the airport on time discovered actions that retrieved the necessary information about times and transport. In each case, the predictive brain simply factored in the availability of reliable internal and external operations and resources as it generated the action policies that would best minimize error in the pursuit of those goals.

We would not expect Mego the orangutan to make good use of information that minimizes uncertainty about global markets, airport arrival times, or even the address of MoMA. But Mego can use a simple stick to gauge the depth of the water. The key difference here, we now see, is not one of bedrock strategy so much as the depth and nature of Mego's understanding of her world. Some understandings, like our own, range over relatively long time scales and more intuitively "abstract" states of affairs. Armed with those kinds of models or understanding, we humans can discover and implement complex sequences of epistemic actions extending over long periods of time.

But whether it is Mego crossing a river, a human being designing a new kitchen, or a group of humans imagining the Large Hadron Collider, the ultimate driving force is the same. It is the biological imperative to resolve critical uncertainties in ways that move us closer to our goals. As this process unfolds, the possibilities provided by some of these external items and resources sometimes become highly trusted, automatically deployed, and deeply woven into our daily lives, turning them into true cognitive extensions.

It is unlikely that the Large Hadron Collider forms part of the machinery of any individual mind. Otto's notebook and my own constantly carried smartphone fare much better in this regard and should be counted as genuine cognitive extensions. In between lie a vast swath of resources (such as the occasionally flaky GPS system in my car) that are less trusted, demand more attention, and are (as a result) more loosely woven into my daily life.

Mind as Brain—Redux

It might be worried that simply by affirming the key role of the brain in selecting and assembling the right set of resources at the right time, we have (contrary to the spirit of the extended mind hypothesis) reselected the brain as the locus of all the truly important activity. If the brain is indeed the primary organ of recruitment, doesn't that also make it the seat of mind and cognition?

The short answer is no. The brain's great skill is that of weaving the right web of resources at the right time, so some problem gets solved. But that doesn't make the newly woven resources merely optional or peripheral. Indeed, in many cases the naked brain simply could not perform the right operations.

So crediting the brain with a primary role in identifying and exploiting a wider web of resources in no way implies that all the credit for the subsequent problem-solving activity belongs to the brain. It is more like the case where a canny leader recruits a crack team of advisors. We should give the canny leader credit for that. But we should not then credit the leader alone with all the nuanced and effective policy visions that then emerge!

The correct response to the worry is thus to firmly distinguish the process of recruitment (the selection of the right resources at the right time) from problem solving that then relies upon the recruited array of resources. The recruited array then acts, together with the biological brain, as a problem-solving "machine" in its own right. This is not unlike the case where the leader then works together with the well-chosen team of advisors. But that example is complicated by the fact that multiple distinct minds are involved. The situation is perhaps even more akin to using one tool (the brain) to make another tool (the larger system), or to the role of a boot program in starting up a computer. We use one set of cognitive processes (the brain-bound ones that serve recruitment) to assemble another cognitive process—a larger problem-solving array comprising a potent mixture of biological and nonbiological resources. Once it is assembled, it is the larger array that solves the problem. The architect looks to the app, scribbles on the sketchpad, and retrieves information from onboard memory, in a complex dance that (all being well) solves the puzzle at hand.

In closing, it is worth noting that there is at least one way of understanding this process of recruitment and exploitation that must be avoided if we are to make room for the idea of extended minds. What must be avoided is the idea of recruitment as itself effortful and deliberative. Here too, there is a disanalogy with the example of the leader recruiting a team.

Instead, my top-level goals act more like catalysts, setting off sequences of actions that call upon external epistemic aids to minimize future prediction error. In this way, the weaving in of the right external resources (in cases of true cognitive extension) "just happens"—it is not the result of effortful deliberation. Otto looks at the notebook because his top-level desire to go to MoMA sets off a processing cascade that recruits a certain combination of inner and outer operations (reaching for the notebook, reading off the address) without the need for further reflection or deliberation. Where minds extend, external epistemic actions should arise and dissolve as fluently and effortlessly as their internal counterparts.

•

I've argued that extended minds arise because predictive brains are naturally expert at exploiting opportunities to use information-gathering action loops to help them achieve their goals. But have we seen, beyond reasonable doubt, that human minds are best understood as extended minds? I am no longer sure that even the best scientific considerations will settle this issue. But there is progress. We now have a much better appreciation of the core principles that allow predictive brains to select actions that bring forth good information using whatever resources are available. This shows exactly how it is that our constructed worlds can sometimes take over, transform, and augment functions once carried out by our brains.

I continue to believe that as the resulting weave between brain, body, and external resources tightens, it becomes less and less productive to think of mind as something locked neatly behind the barriers of skin and skull. But this is at least in part an ethical choice. As the opening example of Tabitha Goldstaub was meant to illustrate, shrinking the individual mind down to the size of their brain-bound processing alone can be every bit as unhelpful as identifying the sporting abilities of a prosthetically enabled athlete with those of the biobody alone. Seen in that light, radical internalism of the "you are just your brain" variety looks retrograde and unhelpful. Yet the alternative option (extending the mind) still feels a little awkward, even to me, and many remain unconvinced.

The new perspective on offer should help us see this debate in a different way. Brains are prediction machines that invoke external resources as easily (and for the same reasons) as they engage practical actions and activate different aspects of their own inner circuitry. As predictive processing unfolds, human experience and human thinking are orchestrated from the inside by neuronal activity and the dense network of brain-body interactions, and from the outside by the highly structured social and technological worlds in which we live and act. This creates a circular causal web in which mind is—at the very least—constantly porous to body and world.

7

HACKING THE PREDICTION MACHINE

WE HAVE seen how the prediction engines in our brain help make us who and what we are. They are not the whole story—indeed, their main role is to enable us to act in ways that engage the complex physical and social worlds in which we live and work. But human experience is shaped by the flow of predictions, so that our every waking moment reflects not just what is coming in from the outside world, but what our brain expected to be coming in. Experience happens only when these forces collide. This, we have seen, opens new doors for treatments and medical interventions. It also suggests ways in which we might begin to take control of our own experiences—ways to "hack our own predictive minds."

In this chapter, we examine some of this emerging landscape. Many familiar hacks can now be seen in a new light—these include making careful use of self-directed language (for example, in practices of self-affirmation), engaging in talk therapy, and appreciating the power of ceremony and ritual. Other hacks we will consider include the deliberate use of placebos and the controlled use of psychedelic drugs. At the far horizons, possible hacks may include the use of virtual realities as a potent new means of relieving pain.

Expecting Relief

Since experience is always shaped by our own expectations, there is an opportunity to improve our lives by altering some of those expectations, and the confidence with which they are held. For as we have seen again and again, it is only confident predictions (even if they are ones hidden from conscious view) that get to exert a real grip on the shape of human experience.

Such confidence can have many sources, and some of those sources involve cultural settings and mental habits. For example, even though I am fully aware that there is no fundamental causal link between wearing a clean, well-pressed white coat and being a good doctor, my brain (courtesy of both past experience and media depictions) responds with confidence to the well-pressed white coat, throwing that information into the pot when deciding how much credence to give to the doctor's assertion that the medication on offer will help me feel better. This means that presentation and ritual can play a potent role in enabling my brain to achieve the kinds of confidence in treatment that are known to play an important role in ensuring the efficacy of certain treatments, especially (though not exclusively) those that aim to relieve pain and anxiety, or to reduce fatigue.

A host of controlled experiments (some of which we sampled in more detail in Chapter 2) have demonstrated that confident expectations of imminent pain relief (from an administered pill or a sham surgical procedure) are often sufficient to cause real and substantial relief. Similar results obtain for nausea, anxiety, immune, hormonal, and respiratory conditions, as well as for migraine relief, lower back pain, and seasonal allergies. Looking outside of the medical context, athletes showed improved performance when hooked up to what they thought were "pure oxygen" delivery systems when in fact they were simply delivering normal air. Similar effects were seen in runners who were led to believe they had been given an injection of a known (and banned) drug that increases red blood cell production. Those led to believe they were "enhanced" were able to improve on their past best performance by 1.5 percent, despite the injection containing only an inert saline solution. Obviously, even small performance improvements such as these mean a lot in competitive sports.

Confidence in a given intervention reflects our confidence in the person (and the larger establishment) offering it, but also the nature of the intervention itself. Injections and surgeries, being considered relatively "powerful" interventions, clearly demonstrate this effect. Patients suffering from various forms of knee (and shoulder) pain were found to improve significantly following "placebo surgery" in which the patient is told they have been given an arthroscopic repair but in fact received only sufficient "surgery" to induce a few incision marks on the area. Remarkably, patients receiving the placebo surgery reported similar amounts of relief as those undergoing normal surgery. Such patients reported substantially more relief than those receiving other, less apparently "serious" placebo interventions such as pills or coaching. This makes sense once we appreciate that the efficacy of the placebo varies with the expected potency of the intervention.

The Strange Case of the Honest Placebo

A fascinating range of cases involves the use of "honest placebos." In these cases, potent predictions of relief can still be activated despite the person knowing perfectly well that there is no standard or clinically active ingredient present.

Honest (or "open-label") placebos have proven effective in cases ranging from irritable bowel syndrome (IBS) to cancer-related fatigue. In one 2010 study, Harvard Professor of Medicine Ted Kaptchuk gave an honest placebo to eighty patients suffering from irritable bowel syndrome and found clinically significant improvements in 59 percent (against 35 percent in a control group), commenting in a later interview that "Not only did we make it absolutely clear that these pills had no active ingredient and were made from inert substances, but we actually had 'placebo' printed on the bottle. . . . We told the patients that they didn't have to even believe in the placebo effect. Just takc thc pills."

It got better. The patients taking the honest placebo (Fig. 7.1) doubled their rate of improvement, equaling the performance of two prominent (active) IBS medications. In another set of studies published in 2016, eighty-three patients with chronic lower back pain were assigned to two groups, one of which continued their medications as before, while the other group was switched to a clearly presented honest placebo. Based on before-and-after questionnaires, those in the former group without intervention reported a 9 percent reduction in usual pain, a 16 percent reduction in maximum pain, and no reduction in disability. But those in the latter (honest placebo) group reported a 30 percent reduction in both usual and maximum pain and—perhaps most significantly of all for daily purposes—a 29 percent drop in experienced disability.

The power of the honest placebo was further underlined in

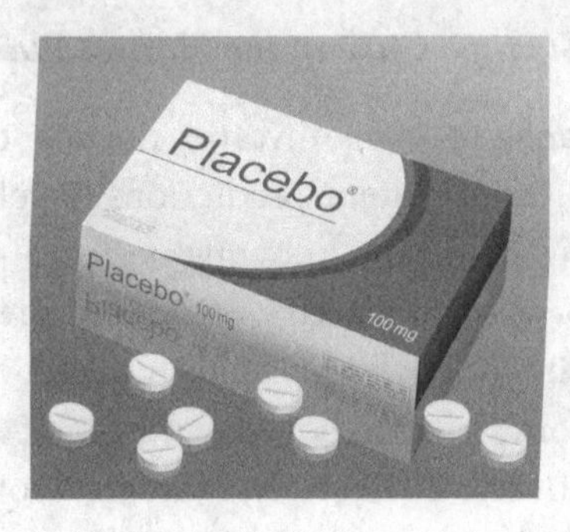

Fig. 7.1 Honest placebos tap into our unconscious expectations.

a 2019 study of patients suffering from cancer-related fatigue (CRF). All subjects were authoritatively informed that the pills they were given contained no active ingredients, and hence ought not be effective in reducing CRF. Nonetheless, the results were striking, leading the authors to conclude that "even when administered openly, placebos improve CRF in cancer survivors."

Honest placebos appear to work by activating subterranean expectations through superficial indicators of reliability and efficacy such as good packaging and professional presentation (foil and blister packs, familiar font, size and uniformity of the pills, and so on). This is because—as we have seen—the bulk of the brain's prediction empire is nonconscious. That leaves it free to respond to quite superficial indicators such as familiar packaging and delivery by those authoritative people in white coats. Such ceremonial features cause the prediction machinery to start to anticipate symptomatic relief despite our conscious belief that no clinically active substance is being administered.

Training Your Placebo

Contemporary thinking about placebo effects is often traced to a wartime incident in Italy when a physician observed a nurse, responding to a shortage of real painkilling drugs, administer occasional saline injections instead of a scheduled injection of morphine. In multiple controlled studies, it has been shown that such "dose-extending placebo" regimes work. The story here is an interesting one. Repeated administration of the actual (clinically effective) drug seems to teach the brain-body system to predict a very specific cascade of pain-relieving responses, many of which can later be re-created or approximated by the patient's own endogenous opioid system—the bodily chemical factory that delivers potent painkilling effects in (for example) life-or-death situations where immediate action, despite serious injury, is required.

A case study involving Parkinson's disease highlights the power of this kind of physiological learning. Patients given a saline injection after repeated injections of the anti-Parkinson's drug apomorphine show apomorphine-like responses. However, no such responses occur if the placebo is given without prior experience of the actual drug. It seems that in these cases experience with the effects of taking the real drug have managed to teach the body how to respond, and the body can then mount those responses for itself when encouraged (by the placebo) to do so. After just four normally spaced genuine apomorphine drug administrations, the response to the placebo was as large as to the real drug. The same broad profile has been seen for other drugs, including the use of placebos to imitate the painkilling effects of aspirin and ketorolac. By using this kind of mixed "dose-extending" therapy, it becomes possible to train precise and effective placebo responses. That approach could potentially be used to improve sports performances in a rather sneaky manner. By training athletes using a performance-enhancing product, and then administering a fake version of the product when they are in competition mode, it may be possible to secure at least some of the benefits of the actual drug while doing so in ways that avoid the possibility of detection.

Next, consider some recent work on improving patient tolerance of statins. These drugs have shown themselves to be a useful part of our ongoing medical efforts to combat heart disease, but adherence is poor due in part to a common misperception that the drugs often have unpleasant side effects involving muscle pain. The best evidence suggests that although these effects can sometimes occur, they are nowhere near as common as many of us believe. It looks as if a lot of us are simply giving up on statins rather too easily, blaming normal aches and pains on the drugs, and perhaps even developing or amplifying such symptoms purely because we consciously or unconsciously started to predict them. This is an example of the placebo effect working in reverse—the nocebo effect, in which expected pain or discomfort acts as another form of self-fulfilling prophecy. It is also a case of expectation contagion—we are led by others to expect the worst, and our bodies then do their best to bring the worst about.

There is, however, a gene variant that has been shown to be associated with the development of statin-related muscle pain and that involves a distinct physiological response to certain statin regimens. This allows for a more personalized approach in which genetic screening identifies those most at risk of statin-related muscle pain. This is good news for those with the gene variant, since they can be offered other statins or alternative treatments.

But here comes the interesting part. Being informed that you do not have the risk-amplifying gene variant was *itself* found to improve tolerance. Simply being told that you are not genetically at enhanced risk of developing statin-related muscle pain itself helps counteract the unhelpful expectations that can sometimes induce such pain. An unexpected benefit of the contemporary move toward precision, or personalized, medicine may thus come directly from the increased confidence we will feel knowing that treatments are highly tailored to our own unique situation. We benefit both from the better targeting that precision medicine offers, and from our own increased confidence in the efficacy of such treatments. This led one of the researchers to comment that "This concept of using precision medicine to address the psychology of how patients feel about drugs might be a winning combination."

In the end, it looks like anything that can be done to increase our confidence in an intervention, procedure, or outcome is likely to have real benefits. This could simply mean trusting a certain doctor, or hospital, or responding to details of packaging and presentation. Much that was previously dismissed as "mere ritual" may now fall into place as part and parcel of how to treat human beings whose expectations of pain and relief are themselves an important part of the causal matrix that delivers their own lived experience. Such effects are already (implicitly or explicitly) understood by medical practitioners of many stripes, as well as sports coaches, life coaches, politicians, teachers, advertising companies, sales personnel, and pretty much anyone who ever needs to deal with other human beings. But understanding them as flowing directly from the normal functioning of the predictive brain paves the way for an evidence-led approach that recognizes both the power and limitations of such effects. Of special interest, as this science develops, will be research that helps reveal why some individuals seem more able than others to benefit from placebo-style interventions and, more generally, from practices and interventions designed to alter their own (mostly unconscious) predictions. Such individuals are experts at what has been dubbed "phenomenological control"—the capacity to exert a kind of unconscious control over the shape of their own experience.

Easing the Pain with Virtual Reality

It is not just pills, potions, and sham procedures that can bring relief. Hooking subjects up to soothing music and rich virtual worlds can be a potent pain reliever too. This has been elegantly shown in recent work by Luana Colloca, a University of Maryland researcher specializing in the neurobiology of pain, with a special focus on placebo and nocebo effects. Virtual reality, Colloca and colleagues show, provides yet another promising means to "hack the predictive brain."

In one of the many VR worlds used in her experiments, a swarm of pulsing and undulating jellyfish float across the field of vision. Under these conditions, heat-pad stimulations applied to a subject's arm were used as a way of testing their ability to cope with increasingly painful intensities. Subjects' abilities to tolerate higher heats were greatly improved when the heat was applied while they were immersed in the virtual oceanic world. Unsurprisingly, opioid treatments were robustly successful too. But opioid treatments combined with VR resulted in even greater reductions in pain than opioid treatments alone. Soothing virtual reality scenes involving coastal walks have been used in dentistry, substantially reducing both experienced and subsequently recollected pain.

VR treatment has also been used successfully in patients with acute burns, enabling them better to tolerate the changing of wound dressings on the burns, and in patients with phantom limb pain. In patients suffering pain from burns, subjective relief using a winter-scene VR program called SnowWorld was similar to that obtained by the use of intravenous opioids. These subjective reports were further borne out by brain-scan data showing significant reductions of neural activity in key pain-processing areas. Reductions in experienced pain have also been found in some controlled studies using various forms of music therapy. But here, the results are mixed and often conflicting. Overall, however, there seems to be a potential role for "combination" therapies in which standard treatments (such as the use of opioids) are combined with other forms of intervention, such as the use of VR, to deliver enhanced benefits.

In 2022, *The New York Times* reported that "The V.R. segment in health care alone, which according to some estimates is already valued at billions of dollars, is expected to grow by multiples of that in the next few years, with researchers seeing potential for it to help with everything from anxiety and depression to rehabilitation after strokes." In 2021, The U.S. Food and Drug Administration authorized the marketing of a VR treatment of chronic pain. Originally called EaseVRx (but now marketed as RelieVRx) , this was a prescription-only virtual reality treatment meant to be combined with other methods such as cognitive behavioral therapy. It included a "Breathing Amplifier" attached to the headset for use in deep-breathing exercises. The primary target of EaseVRx was chronic lower back pain, a condition that affects millions of adults in the U.S. The current incarnation, RelieVRx, involves daily seven-minute immersive VR experiences spread over eight weeks. It was designated as a Breakthrough Device by the FDA, and early results were promising, showing significant reductions in pain relative to controls. This has since been confirmed in a double-blind, randomized, placebo-controlled study.

Why does VR treatment work? The natural but somewhat superficial answer is simply "by distraction"—by prompting us to focus less on the pain and more on the soothing novelties of the floating jellyfish. But predictive processing digs a little deeper, suggesting the shape of the underlying mechanism and making better sense of the importance of the specific content too. The key here is the immersive environment. This—as anyone who has tried contemporary VR will attest—is pretty much impossible to ignore. It is alien and surprising, but never threatening. Such a sensory world acts like a magnet, requiring the brain to scramble to make sense of this new and constantly shifting environment. Increased precision over this new sensory information means decreased precision over other sensory information, including information regarding pain. At the same time, the gentle rise, pulsing, and fall of the jellyfish engage and entrain our own bodily rhythms, altering our breathing and heart rate. Such alterations, as we saw in some detail back in Chapter 4, act as evidence that can further tip the predictive balances in ways that favor a peaceful, relaxed attitude.

The well-chosen VR world is not merely distracting us—it is subtly changing the predictions that build experience. The use of associated breathing and behavioral techniques alongside immersive VR also now falls neatly into place. These all work together (along with explicit coaching) as a coherent nonpharmacological package able to nudge the embodied prediction machine in ways suggestive of calm and relief.

Cautions and Tangles

Nudging the prediction machinery along in all these various ways is a very good trick. It is worth stressing, however, that there are multiple mechanisms operative in disease and health, many of which are simply not impacted by placebos and related interventions. Treating cancer-related fatigue is not the same as treating cancer, and no amount of carefully sculpted self-expectation can replace a splint when healing a broken arm. In the case of Parkinson's, placebo drugs reduce pain and muscle rigidity but—and this cannot be overemphasized—they do not seem in any way to affect the process of neuronal degeneration that underlies the steady progression of the disease. Similarly, no placebo affects the bacteria that cause pneumonia. There is, as Fabrizio Benedetti (a leading researcher in the field of placebo effects) notes, no evidence that placebo responses occur for many classes of drugs including antiplatelets and anticoagulants. Benedetti also warns us against the very real danger that hard science could here give an unwelcome (and potentially dangerous) boost to pseudoscience. This could happen if the demonstrable (but limited) efficacy of placebos were to be hijacked as "evidence" of the true causal potency of specific alternative therapies.

Appreciating the potential role of placebos can also lead to personal and ethical tangles. This was brought home to me after my mother was diagnosed with terminal cancer. She was put on a hormone therapy that was briefly very effective in delaying the inevitable, allowing her to celebrate her eightieth birthday with customary style and panache. At some point in her treatment, however, the color of one of the pills that she was taking was altered—from a vibrant pink to a dull blue. My mother insisted that the pink pills made her feel much better. Checking the ingredients, we could find no clinically significant difference. The doctors confirmed as much. My mother firmly but gracefully demurred. But try as we all might, no pink versions could be sourced.

This put me in a tough spot. I knew that placebo effects were real, and potentially helpful against fatigue and anxiety, and I knew that for her, the pink ones really gave her confidence. Yet the best way forward, it seemed to me, was to firmly suppress this thinking on my part, and try to convince her of the equal efficacy of the blue pill, thereby (I hoped) rendering it indeed equally efficacious. I also genuinely believed that in some deep sense the color really ought not to matter. Blue really should be as good as pink! And it would be, if only the right set of expectations of relief could be engaged.

To this day, I'm not sure how I should have responded. It was (and remains) a tangle. Notably, it wouldn't have been any different even if I'd been the one taking the medication. I think we have a long way to go before a solid picture emerges of how best to leverage potentially powerful placebo effects in an honest, helpful, and evidence-led way.

The Power of Self-Affirmation

No such complications affect our next hack for the predictive brain. Just as placebos and rituals can impact the deep prediction engines that sculpt human experience, so too can verbal interventions of various kinds. Cases in point include the strings of comforting words uttered by someone well trusted (for example, in the context of talk therapy), but also the words we ourselves use, either actually uttered or in inner monologues, to frame our own thinking. In these and many other ways, the careful use of language has the capacity to reach into the heart of the experience machine.

A well-studied example is the positive, performance-enhancing power role of self-affirmation. The classic self-affirmation procedure used in such studies involved asking participants to write a list of positive characteristics (abilities, skills, or personality traits) they possess, perhaps expanding on one that they consider especially important, and recalling a time it made a difference in their own lives. A control group would be asked to write about someone else's positive characteristics. Both groups are then asked to perform some unrelated but challenging task, such as a math test or a test of spatial reasoning abilities.

Under such conditions, those prior acts of positive self-affirmation have powerful consequences. These consequences were greatest in cases where the subsequent test targets a skill that tends to induce performance anxiety in the participants. This seems to be because the self-affirmation practice pushes back against the negative performance effects brought about by various forms of "stereotype threat"—the tendency we have to perform badly at tasks where we (consciously or unconsciously) predict we will do badly, creating a kind of self-fulfilling prophecy of poor performance.

There are now many examples of this. Completing the prior self-affirmation task almost completely abolished an otherwise striking gender gap between male and female participants in some math and spatial reasoning tasks. The effects can outlast the experimental settings too. In the U.K., self-affirmation training was given to a group of students from socioeconomically deprived backgrounds, and this intervention reduced the difference in exam results (compared with children from more affluent backgrounds in the same class and school) by 62 percent. Similarly, Black students in the U.S. performed a self-affirmation exercise for around fifteen minutes at repeated intervals during their schooling, reducing otherwise prevalent racial differences in exam performances by 40 percent.

Reframing Experience

We can also use words to frame and reframe our own experiences and anxieties. This is another potent tool whose powers and mechanisms can now be better understood. For example, consider that prickly rush of adrenaline so often felt before going onstage or delivering a speech. We can practice attending to that feeling while verbally reframing it as a sign of our own chemical readiness to deliver a good performance. This can lead to more relaxed and fluent behavior. Carefully chosen language can select, enhance, or dampen many parts of the web of neural guesswork that builds human experience. Once again, this is not mere framing but really a deeper effect, impacting the construction of the core experience itself.

Striking examples can also be found in the literature on hunger and satiety. There, beliefs about the composition and caloric value of foods have been shown to influence the extent to which we feel satiated after eating them. In one experiment, well-matched student subjects ate identical breakfast omelets but their expectations of later hunger were subtly manipulated by the experimenters. This was done by (before serving the omelets) showing one group pictures of a large cheese-and-four-egg omelet, while others were shown images suggesting about half those amounts. All were then served an identical intermediate (three-egg) omelet.

The students whose prior experience involved the larger depictions experienced less hunger and chose to eat smaller portions at lunchtime than those who saw the two-egg depiction. Other work has shown similar effects induced by labeling and description (identical 380 calorie milkshakes described as "rich and creamy" versus "light and healthy"). This work also showed that the effects of varying description-driven expectations reached deep, impacting not just experienced hunger but also the secretion of the hormone ghrelin, which helps regulate the way the body uses energy and burns body fat.

Next, consider the approach known variously as "pain reframing" or "pain reprocessing theory." We may often misunderstand the "meaning" of our own pains. What a bodily pain often seems to be saying to us is something like "don't do that, you will damage yourself seriously if you do." This is usually correct in the case of acute pain—the pain you feel, for example, soon after you start accidentally cutting your own finger on the chopping board. Stopping your cutting activity is certainly the right thing to do at that point! But very often, in cases of chronic pain, the pain system itself has been compromised and we should no longer trust in that original meaning.

Experienced pain has been usefully glossed as "an unpleasant sensory and emotional experience associated with, or resembling that associated with, actual or potential tissue damage." What this definition implicitly recognizes, by adding the "resembling" caveat, is that there are indeed multiple ways that the pain system can become compromised so that the inference from "this hurts" to "I shouldn't be doing this" becomes unreliable. This is frequently the case with chronic pain, but the importance of this as an opening for new therapies and interventions has been underappreciated.

Recall from Chapter 2 that pain experiences were traditionally understood as flowing from two main sources. Either a pain was "nociceptive" in origin, meaning that it reflected the kind of nervous system activity associated with actual or threatened tissue damage. Or it was "neuropathic," meaning it was caused by damage or disease affecting the nerve networks themselves. Yet in many cases of chronic pain, neither of these conditions seemed to be clearly met, or met in any way consistent with the degree or nature of the ongoing pain experience. So in 2016, a third category was added to the list. This was "nociplastic" pain, defined as "pain that arises from the abnormal processing of pain signals without any clear evidence of tissue damage or discrete pathology involving the somatosensory [nerve network] system." In other words, real pain experiences rooted not in structural damage, but in anomalous processing. Nociplastic pain is thought to arise when otherwise innocent sensory signals are amplified, or when pain-inhibitory systems are dampened. These are, of course, precisely the kind of effects that can occur at many different levels in predictive processing.

In practice, chronic pain seldom falls solely into one of the three categories (nociceptive, neuropathic, or nociplastic). Instead, there is a continuum of cases and a great many—especially cancer pain and spine pain—seem to involve complex mixtures of all three. What matters for present purposes is not to shoehorn anyone's experience of chronic pain into one of the categories, so much as to recognize that a lot of the disability that comes with chronic pain and injury is linked to that hidden inference—the inference that the felt pain means we ought not to push ourselves any further. In the grip of that inference we shrink our worlds, lending further support to the prediction that we are simply not capable of doing many of the things that would otherwise expand and enrich our lives. Yet in many cases of chronic pain and disability, this inference is mistaken.

We saw some quite dramatic examples of this in Chapter 2, where misfiring patterns of attention were amplifying innocent bodily signals, creating clear experiences of pain despite the lack of any standard medical cause. Instead, it was the misfiring patterns of attending and expectation that brought about the (100 percent real) experiences that afflicted individuals report. In such cases, there is no gross systemic damage to exacerbate—no damage to the affected bodily systems or to the structural integrity of the nerve networks ostensibly reporting on them. Instead, the culprit is an aberrant pattern of neural processing involving misplaced predictions and misguided levels of confidence in those predictions. It is these aberrant predictions and precisions that deliver the perception of pain, paralysis, or even blindness.

Ordinary (nonfunctional) forms of chronic pain and chronic fatigue present a related kind of case, one that is perhaps closer to many of our daily experiences. Here, there may often be some normal medical cause present. But that still does not mean that the pain or fatigue should always be interpreted as an imperative to curtail activities so as to preserve energy or avoid further damage. Instead, it may (for example) reflect a kind of self-maintaining hypersensitivity (sometimes called "central sensitization"). This acquired hypersensitivity maintains the pain experience long after it has ceased to be able to play the kind of damage-limiting role that is its adaptive rationale. A similar profile applies for at least some cases of chronic fatigue.

This is where pain reframing and pain reprocessing can play a helpful role. These interventions normally involve advice, training, and counseling that first expose and then push back against that hidden inference from pain (or fatigue) to activity restriction. All this—when it works—can support the staged reintroduction of previously curtailed activities. But there is an important burden of education here, for without a proper understanding (by both patient and physician) of the mechanisms involved, such attempted interventions can easily misfire. They will misfire if they are seen as suggesting that the pain or fatigue is not real, or is "only in the mind." Instead, the point of the interventions is to push back against the misplaced predictions and precisions (each involving genuine neuronal changes) that are positioning otherwise innocent bodily signals in such a distressing and maladaptive way. To begin to reverse these changes, the idea is to provide new evidence able to drive a different set of predictions. At the same time, careful verbal reframing seeks to destabilize the old ones.

Perhaps the single most crucial element in bringing about positive outcomes involves explaining to an affected individual that their experienced pain or fatigue, although real, need not signify any threat of imminent damage and that their felt inability to perform certain tasks is thus acting more like a cause of disability than a reflection of it. A more detailed version of that kind of reframing might include describing exactly how it is that genuine pain can be caused by misfiring prediction systems. Patients may also be advised to attend to the sensations in as much detail as they can, but not under the label "pain" or "hurt." All this is an attempt to destabilize the role of aberrant attending (aberrant precision weightings) and misfiring predictions, so as to unseat the old inference from pain to incapacity, allowing the formation of a new and more helpful set of self-expectations. In addition, many physicians stress that nociplastic pain is curable, as this sets up a new high-level expectation that can itself play a powerful role in bringing about positive change.

To see how all this works in practice, let's look at a few examples.

Pain Reprocessing Theory

In a short piece in the *British Medical Journal* entitled "Reframing My Chronic Pain," Hannah Vickers shares her own experience of chronic pain, observing:

> The unpredictable nature of pain symptoms can lead to overthinking, overplanning, and avoidance. When I first experienced chronic leg pain I noticed that stairs could be triggering it. As a young, fit individual I began to actively avoid stairs and seek out the lift. This meant that, when presented with stairs as the only option, I often became nervous and then hyper-aware of my legs. When I then did go upstairs, inevitably it would hurt and the experience would reaffirm the idea that stairs were bad for me.

After verbal interventions that challenged the inference that the pains were a reliable indicator of imminent bodily damage, and others that helped her appreciate that they would tend to come and go pretty much regardless of the activities themselves, she was able to pursue a staged return to many daily activities. The most effective verbal interventions here often involve changing the language in which to talk about the pains, avoiding conceptualizing them as reflections of damage or indicators of disability. For example, Hannah later comments:

> I have started to call my daily exercises "training," instead of physio, to reflect this and to relinquish ties to injury or disability. Having these positive aims helps me create a more optimistic landscape to better navigate the hardship of chronic pain.

The fit with predictive processing is tight. Reframing pain by talking and thinking about our feelings and sensations in a slightly different way, we enable our brain to reshape its own prediction landscape. And it is that landscape that, as we have seen again and again, in turn shapes human experience and human action.

What we experience is nothing other than the conclusions of our brain's best inferences—its best guessing at how things are in body and world. If the best guess is that we cannot do very much without further damaging ourselves, the result will be a highly restricted realm of action. But if (thanks to the kinds of training just mentioned) we find ourselves doing a little more than we thought possible, our brains gather evidence that can drive a different inference—one whose conclusion is that we are not as restricted in our actions or likely to experience as much suffering as we previously thought. At that point the pain itself feels less, so doing a little more becomes a little bit easier. This can lead to a kind of virtuous circularity, in which doing more feels less painful or tiring, and that experience adds further weight to the belief that we can indeed do more than we previously thought.

Here, too, immersive VR may have a role to play. Karuna Labs in San Francisco has been exploring ways to use VR to give people experiences of moving their virtual bodies in ways that—due to the pain they tend to experience—they would not normally risk in daily life. They might, for example, increase spinal flexation using an archery game setting. This can then have benefits outside the VR setting, as the VR experience gives the brain reasons to downgrade the predictions of movement-related pain that (as we have seen) are causal factors in many cases of chronic pain.

Escaping Our Own Expectations

Verbal reframing of various kinds is a powerful technique for altering the predictions that shape experience and that determine action. But simple reframing is not always enough. Sometimes our inner models are so inadequate or (as in chronic depression) are so deeply locked-in, that we need not simply to tweak them, but to break and re-form them. This kind of radical overhaul is not easy, but there are ways of bringing it about.

The most obvious way is, of course, simply by new, slow, and perhaps painful learning. If I have no good model for how to drive a car, or play cricket, the best way to get one is to immerse myself (usually with careful guided instruction) in a car-driving or cricket-playing environment so as slowly to train a new set of perception and action routines. But there are also cases where what is needed is not so much to learn a new model as to break the viselike grip of an old one that is no longer serving our purposes.

A few years ago, I came across an unusual (but fascinating) illustration of this kind of model breaking. It involved Max Hawkins, a computer scientist turned artist. His story began when, working as a Google engineer (his dream job) in San Francisco (his dream city), he realized that he had optimized his life to fit his preferences to what he suddenly found to be an alarming degree. He started every day at the stroke of 7 a.m., went to the best coffee shop, then cycled an optimal fifteen-minute route to work. A simple algorithm called GPS Tracker, fed with his data from one week, could predict with great accuracy his whereabouts and movements the next week at the same time of day. This smacked, he felt, of a certain lack of personal autonomy. Despite having fit his life almost exactly to his preferences, he felt trapped—as if he had optimized his life to the point where nothing really surprised him or gave him much real pleasure.

In response, Hawkins progressively outsourced his decisions and choices in a very unusual way. Using his computer engineering skills, he built a wider web of technology designed precisely to challenge his own behavior-recruiting expectations and force him out of his comfort zone of sunshine and sushi. For two years, he lived his life according to a series of randomization algorithms. A diet generator told him what to eat, an algorithmic travel agent picked out the city where—having gone freelance—he would live for the next two months, a random Spotify playlist provided music for the journey, and a random Facebook event selector told an Uber driver where to take him when he got there. They took him—to mention just a few—to acrobatic-yoga classes in Mumbai, to a goat farm in Slovenia, to the small-town pub of Holy Cross, Iowa, and to an eighth-grade flute recital. Anywhere, he explains, that would break him out of the comfortably predictable rut of the affluent San Franciscan tech worker.

Hawkins's algorithms dictated not just where to go, what to eat, and what leisure activities he should engage in, but even what clothes and hairstyles he should adopt (he ended up needing several wigs). He even sports a chest tattoo selected randomly from images on the web. Hawkins reports finding great fulfillment in multiple unexpected ways and feeling (paradoxically) more present as a person as a result of escaping what he had come to see as the dictatorship of his own preferences and expectations. He talks of escaping the tiny "bubbles" of places to eat and things to do that kept on dragging him back time and time again. This all-out embrace of uncertainty and surprise can seem strange, even paradoxical. For as we have seen, brains like ours are designed to minimize their own present and future errors in prediction. The more unpredictable the environment, the less error gets minimized, often resulting in anxiety, stress, and feelings of loss of control. Yet there is Hawkins, joyfully adding huge doses of the unexpected into his life.

But as we saw back in Chapter 4, simply staying within the bounds of the expected is actually only one part of a much more complex story. For those very same predictive brains were designed to drive mobile, inquisitive creatures like ourselves. Such creatures must also explore and discover new opportunities. The two are consistent since it is often only by exploring local pockets of increased surprise and uncertainty that we discover new ways to meet our needs. Deliberately engineering restricted forms of surprise allows us to balance the need to exploit what we already know and to explore further afield—an achievement formally identical to the way foraging animals balance reaping predictable rewards with engaging in further exploration.

Hawkins was really in the business of structuring his world (via the algorithms) in ways that could probe and stress-test his own deeply entrenched assumptions about who he was and what he liked. This is just another way of exploring more of the space of personal possibility. The methods were extreme, but the general project both familiar and distinctively human. Hawkins was exploring beyond his usual bubble. But he was not in the business of rendering his life truly unpredictable, so much as "differently predictable." After all, it is notable that there was still a set procedure in play, and one that he himself understood (indeed, one that he designed). For example, he knew that the algorithm would send him somewhere new every two months. It would not randomly uproot him at any arbitrary time.

It was probably this higher-level kind of predictability that kept him feeling safe and sane, and that enabled him to gain so much from his experiment. Even when we engineer our worlds to deliver surprise and enable learning, we do so in ways that limit surprise itself in mostly predictable ways—just as a roller coaster delivers a set of physical surprises of a well-known and (in that sense) broadly predictable kind. Hawkins also observed how rapidly the strangest and most unlikely situations and places for him became the "new normal." So much so, he said, that he could easily start to imagine the rest of his life in that once alien place. This, I conjecture, is the predictive brain reasserting itself, re-forming aspects of our own high-level self-model to get a grip on the new way of life.

Hawkins's takeaway message was simple: don't let your own preferences become a trap. Yet even in his new lifestyle, he remained trapped—in a good way. His randomizing algorithm simply fulfilled his new (and temporary) prediction that he would experience controlled doses of change and exploration.

Psychedelics and the Self—A Chemical Romance

Hawkins was trying to improve upon what, to most of us, would seem like a pretty good situation. There are many cases, however, where our own hidden expectations are causing us not just boredom but real harm. For some of these kinds of cases, it may be helpful to explore even more radical means of loosening the grip of our own entrenched self-models.

The last decade has seen a growing body of scientific studies that demonstrate the remarkable efficacy of controlled doses of psychedelic drugs such as LSD and psilocybin for a variety of problems including addiction, obsessive-compulsive disorder, PTSD, treatment-resistant depression, and as a way of coping with "existential distress" in end-of-life care. Moreover, new research suggests that positive outcomes can be shared in nonclinical populations too and may include greater feelings of connectedness with nature and other people, as well as improved ecological awareness and reduced anxiety.

Quite recently, an important body of work has started to take shape linking these benefits to quite specific alterations to patterns of information flow in the predictive brain. Leading psychedelics researcher Robin Carhart-Harris describes the benefits of these chemical interventions as "shaking the snow-globe." The idea here is that the drugs can help jolt the mind out of entrenched negative patterns, making it more flexible and open. In the cases of many affective disorders such as severe and chronic depression, the simple fact of temporarily experiencing the world very differently can be liberating. What had seemed like deep, immovable facts about how things are (and how you are) release their stranglehold, allowing other ways of seeing the world and "being you" to emerge.

Intriguingly, somewhat similar benefits can sometimes be obtained using virtual reality settings that immerse the user in a dazzling world of intense visual images. Brennan Spiegel, a leading proponent of what is becoming known as Virtual Therapeutics, likens the effect explicitly to that of psychedelics such as psilocybin, referring to the intense VR experiences as "cyberdelics." Forcing the brain to try to get to grips with a vivid new world may be yet another way of "shaking the snow-globe" and allowing new self-understandings to take shape.

Such effects are more pronounced, however, with the use of drugs. Under their influence you can still be you, while experiencing things in ways that seem strangely discontinuous with your own previous preoccupations and sense of self. When this works well, there is suddenly room to rediscover the old romance—to start to feel at home again with yourself and the world. The writer, philosopher, and psychedelic pioneer Aldous Huxley, in this 1954 account, *The Doors of Perception*, put it like this:

> To be shaken out of the ruts of ordinary perception, to be shown for a few timeless hours the outer and inner world, not as they appear to an animal obsessed with survival or to a human being obsessed with words and notions . . . this is an experience of inestimable value.

Huxley, strongly influenced by the poet William Blake, thought of ordinary perception as a kind of "reducing valve" that blocked out most of the true structure of the world, delivering merely the "trickle" most helpful for survival.

Consequently, a recurring theme in *The Doors of Perception* is that of sensory blockage, unblockage, and subsequent revelation.

Predictive processing both supports and subverts that image. It supports it, insofar as it agrees that experience is shaped by our own high-level beliefs (predictions) about ourselves and the world. Relaxing the grip of certain high-level beliefs may indeed allow more of the incoming signal from the world to (in a certain sense) speak for itself. However, there is no way to simply release the "reducing valve" and let it all in—no way to just unlock the doors of perception and let the world freely impose itself on the mind. It is only by bringing the sensory flux into contact with predictions and expectations that experience takes shape at all. But we can reduce the effects of certain high-level predictions and self-expectations, creating room for others to form and take root.

Psychedelics, Carhart-Harris suggests, may do just that. They may help relax the grip of our existing model of who we are, what we will do, and what is most meaningful in our life. We can then experience the world, ourselves, and others in new and liberating ways. Much of the distinctive experiential feel (the "phenomenology") of psychedelics may be explained in this broad fashion. To see how, it helps to look at the specific way these substances alter the precision-weighted balances that characterize the predictive mind.

Loosening the Grip

Psychedelic drugs exert their strongest effects at higher levels of cortical processing. This suggests that they are likely to be impacting higher-level (increasingly abstract) elements of the predictive model of self and world. Those elements might concern who we are, what we want, and how we conceive of reality itself rather than more concrete matters of color, taste, and shape. Psychedelics also dampen the activity that communicates predictions between different neuronal populations. And they dampen the activity of key brain networks (notably the "default mode" or "resting state" networks) that are known to be especially active during self-centric thinking and rumination. Seminal neuroimaging (fMRI) work by Carhart-Harris and colleagues has also shown that the changes in conscious experience that psychedelics promote are associated with decreases, rather than increases, of brain activity—something that, I strongly suspect, would have shocked Huxley with his image of (essentially) opening the floodgates.

Putting this all together leads Carhart-Harris to what he playfully dubs the REBUS (RElaxed Beliefs Under pSychedelics) model. According to REBUS, psychedelics relax the grip of higher-level expectations concerning self and world, allowing different flows of information to emerge and interconnect in new ways: ways less constrained by our ingrained top-level expectations. Such effects are dose dependent. At low doses, perceptual—and especially visual—effects are dominant. But increased doses impact functioning at ever higher levels and can then lead to classic psychedelic experiences such as loss of the sense of self (ego dissolution) and feelings of oneness with nature.

At these high doses, Carhart-Harris speculates, the drugs seem to be reducing the precision-weighting on highly abstract top-level self-predictions, thus releasing bottom-up sensory and bodily information, and allowing such information to play a larger role. This can be liberating in much the same way as finding yourself in a foreign country where no one expects you to behave in the way you do back home. The process is somewhat akin to heating up (annealing) a metal to induce a temporary state of greater plasticity during which new forms can be explored, some of which may later stabilize. The later stability would correspond to the longer-term benefits associated with successful interventions of this kind—a bit like returning home but still seeing things in a new and helpful way.

But there is risk in entering such a hot, malleable state. It is not uncommon, for example, for experience with psychedelic compounds to lead to the too easy adoption of pseudoscientific and supernatural beliefs. This could occur as a direct result of the sudden awareness of the extent to which how things appear to us is a product of our own minds, and as a kind of rationalizing response to the feelings of great and significant interconnectedness mentioned earlier. The best remedy for this is to be well informed—as individuals and as a society—about the action of psychedelics on the brain. Properly informed, we can engineer psychedelic experiences, in controlled environments, that include careful training and follow-ups that promote the most helpful kinds of insight while reducing the danger of supernatural interpretations.

The potential therapeutic benefits of relaxing the hold of top-level self-expectations are strikingly obvious in the case of treatment-resistant depression, anxiety, PTSD, and other affective disorders. But there may also be benefits for anyone (and I suspect this is everyone) who might learn by experiencing their world—even for a short while—in a less entrenched and ego-driven way. Such benefits would be especially marked, as my colleague Anil Seth has pointed out, for anyone prone to the dangerous illusion that their current way of seeing the world and experiencing themselves reflects some fixed or fully objective truth.

Meditation and the Control of Attention

Properly used, psychedelic drugs offer a way to step back from our ordinary daily doubts and self-concerns, providing what has been described as a "holiday from the self." This is also one of the key effects of meditation, a practice that likewise quiets the ego, as evidenced both by verbal reports and by dampened neuronal responses in areas (such as the default mode network) associated with introspective self-consciousness—the same areas in which activity was seen to be dampened by the action of the psychedelic drugs.

It is unsurprising then that the meditative route to these effects is itself now being understood using the tools and constructs of predictive processing. Focused-attention meditation provides a good example. In focused-attention meditation, practitioners learn to maintain attention on a single object such as the breath. In predictive processing terms, upping the precision on that sole reliable object inevitably results in dropping the precision assigned to all other states, effectively down-weighting all the rest of the information flowing in from our senses. Once this skill is acquired, thoughts, memories, and sensations can also arise without capturing attention. This means they can be experienced in a way that is helpfully disengaged from our normal tendencies to react and respond. A bodily itch or a disturbing thought may still arise, but it is not experienced as an immediate call to action, such as scratching or rumination.

By clamping attention on to an unfolding present moment (such as the breath), we also temporarily shrink the time horizon of predictive processing. This implies a kind of freezing of longer-term anticipatory processes, preventing the kinds of counterfactual "looking ahead" that play such a large role (as we saw in Chapter 6) in daily behavior. This means that even internal "information foraging" (purposeful explorations of our own memory, for example) can be put on hold. This is awareness with minimal counterfactual and temporal depth.

Much more research is needed before we can be confident of the nature and significance of the various brain changes that result from long, sustained practice of this kind. But at a very general level, it now seems that what the experienced meditator has acquired is a new lever for the control of previously automatic aspects of their own neural activity. This is what it really means to learn to "control attention." It is to gain better control over the precision-weighting performances of our own brain. This, in turn, enables the experienced meditator to enforce a kind of temporary distancing from their own current situation and ongoing concerns. Sensations, fears, hopes, and memories can then be experienced without engaging the usual wheels of judgment, rumination, and calls to action. Many contemplative traditions and structured physical practices are neatly positioned to train and promote just such forms of attentional control, often using bodily awareness of breath, posture, and muscular tension as a tool. By training attention and bodily awareness, such practices can teach us ways to exert enhanced control over our own thoughts, feelings, and experiences.

•

In sum, there are many ways to nudge, prompt, and manipulate our own predictive brains. These include the careful use of self-directed language, talk therapies, pain reprocessing strategies, meditation, psychedelic drugs, and (as we saw in the case of honest placebos) packaging, presentation, and ritual. Work on the embodied predictive brain provides the first framework with the power to understand, link, and triangulate these superficially very different factors and forces in a truly principled way. Perhaps in the future we will all benefit from carefully tailored training regimes that deliver greater control over the prediction machinery that so powerfully shapes the way we experience our bodies, selves, and worlds.

Predictive brains are the active constructors of every facet of human experience. The better we understand that process the more we may sculpt and leverage it in ways that promote flourishing and success.

Conclusions:

Ecologies of Prediction, Porous to the World

WE ARE what predictive brains build. If predictive processing lives up to its promise as a unifying picture of mind and its place in nature, we will need to think about ourselves, our worlds, and our actions in new ways. We will need to appreciate, first and foremost, that nothing in human experience comes raw or unfiltered. Instead, everything—from the most basic sensations of heat and pain through to the most exotic experiences of selfhood, ego dissolution, and oneness with the universe—is a construct arising at the many meeting points of predictions and sensory evidence.

At those busy meeting points, nothing is passive. Our brains do not simply sit there waiting for sensory stimulations to arrive. Instead, they are buzzing proactive systems that constantly anticipate signals from the body and from the world. These are the brains of embodied agents, elegantly designed for action in the world. By moving our eyes, heads, and limbs we seek out the sensory signals that will both test and (usually) confirm our predictions. Experience takes shape as predictions of our own sensory inputs are tested, refined, and challenged in these ways.

Prediction error signals result when current predictions fail to fit those waves of incoming evidence. It is these prediction errors that now do much of the work, keeping us honest, keeping us more-or-less in touch with how things are in body and world. This flip, though subtle, is important. It means that although we are open to correction, the corrective process itself reflects the shape of our own prior expectations. The stronger our prior expectations (the higher their "precision") the less impact incoming counterevidence will have on what we see or otherwise perceptually experience. It is this precision-weighted balancing act that seems to be compromised in many forms of mental illness, functional disorder, and psychosis. There are also many different ways the balancing act can be performed, and these plausibly correspond to the wide sweep of neurodiverse ways of being in the world.

We must not forget that prediction, as it features in these accounts, is not just (or even mostly) a matter of what someone might say or think that they predict or expect. In case after case, the predictions that sculpt and inform human experience have been shown to be invisible or shrouded from the conscious mind. When we see a back-lit hollow mask as a forward-facing face, we are not aware that this is an illusion caused by our brain's deep expectation of convex face-forms. Ditto for the many other predictions that push and pull our experience—not just of the outer world but also of our own medical symptoms and changing bodily states. The bulk of the predictions and expectations that determine the shape of human experience are hidden from our own view and seldom formulated in words. But despite this, our brains are teeming with these active predictions and they impact everything we see, feel, hear, and touch.

But predictive processing is more than a new picture of perception. It is a new picture of action too. Most excitingly, it offers the first fully unified treatment of perception and action. It displays them as co-constructed around the common goal of minimizing error in the prediction of sensory states. To perceive is to find the predictions that best fit the sensory evidence. To act is to alter the world to bring it into line with some of those predictions. These are complementary means of dealing with prediction error, and they work together, each constantly influencing and being influenced by the other.

It is this deep reciprocity between prediction and action that positions predictive brains as the perfect internal organs for the creation of extended minds—minds enhanced and augmented by the use of tools, technologies, and the complex social worlds in which we live and work. Extended minds are possible because predictive brains automatically seek out actions that will improve our states of information, reducing uncertainty as we approach our goals (highly predicted future states). When such actions become parts of habit systems that call upon resources that are robustly available, trusted, and fully woven into our daily ways of dealing with the world, we become creatures whose effective cognitive apparatus exceeds that of the biological brain alone.

This is a satisfying picture. It reveals us as creatures that are truly of our worlds: creatures whose percepts and actions all flow from a single source. That source is the ongoing attempt to minimize (precision-weighted) prediction error as we go about our lives. That impressionistic gloss is, moreover, firmly grounded in an account of the core computational operations involved, and the way those operations may be realized by brain structures and processes. But the picture, although attractive, remains in several ways incomplete.

Many will think that the biggest unsolved puzzle concerns the origins of qualitative conscious experience itself. This is a murky topic that I have deliberately (apart from a few sketchy remarks in the Interlude) avoided confronting head-on. But the title of the book is *The Experience Machine*, and I stand (nervously) by my title! I believe that understanding the way predictions of many kinds alter and adjust the shape of experience is our biggest clue yet to what experience really is, and how it comes to be. In fact, I suspect that by doing a whole lot of science of this kind we will slowly dissolve the so-called hard problem of explaining the nature and possibility of qualitative experience itself.

At the very least, predictive processing tells us a lot about what might be described as "basic sentience." Basic sentience occurs whenever a creature encounters its world as a meaningful arena for action—when it treats different states of that world as attractive or unattractive, as providing or foreclosing opportunities. By bringing work on the predictive brain together with the role of actions, error dynamics, and internal bodily signals, we laid out what seems to me to be the rough shape of a theory of basic sentience.

But on top of all that, in the human case, are laid multiple more exotic forms of knowing and representing ourselves and our worlds. These include the use of structured public language, and the systems of writing and cultural transmission that enable us to train new minds using the collective wisdom of previous generations. These abilities have led us to become a strangely self-reflective species and have plausibly played a major role in enabling us to harbor thoughts and ideas about our own minds and their contents. I suspect that once basic sentience is explained what remains is then mostly a set of misleading intuitions—cognitive illusions rooted in our peculiar, perhaps linguistically inculcated, abilities of self-reflection.

Material symbolic culture poses important unresolved challenges. We need a much better understanding of the many ways material culture and the social and physical environment alter and impact patterns of prediction and action. We explored the fundamental merging of epistemic and pragmatic concerns enabled by the predictive brain. But in us humans, this has slowly synergized with an additional suite of capacities to use complex spoken sentences, make sketches, draw diagrams, and write things down. Such capacities all involve the use of representations—external encodings whose role is to represent useful information. The origins of our species' distinctive abilities in these regards remain uncertain. But however they arose, these skills at external encoding provided powerful new tools for minimizing expected future prediction error, at time scales that would defeat most other animals.

As we engage with these complex sociocultural worlds, it remains unclear exactly how our conscious expectations interact with multiple forms of nonconscious prediction. There is no doubt that they do. We saw, for example, that conscious expectations make a systematic difference to experienced pain. But since there are so many nonconscious predictions and expectations also in play, we need a much better understanding of how and why explicit conscious predictions sometimes play greater or lesser roles in sculpting experience. Future work should target the detailed nature of this influence. Improving our understanding of this process is clearly essential if we are to become systematic and effective hackers of our own predictive brains.

Despite these shortfalls, we now have good grounds to believe that all human experience is built using the brain's best predictions. This suggests a surprisingly large space for individual variation, since different people will always bring different predictions, and different precision-weighted balancing acts, to the task of perceiving and acting in their worlds. There will be differences too in the amount of control different individuals exert (and learn to exert) over those balances. Understanding the nature and effects of differing balances within the experience machine also locates neurotypical and atypical forms of human experience within a single, unifying framework, in ways that have significant implications for psychiatry, medicine, and clinical practice.

On the horizon we glimpse a new evidence-based science—one linking the ebb and flow of predictions and precision-weighted error signals to all the many large and subtle variations in how we experience and act upon our worlds. Once we have that science in hand, interventions upon experience should become much more reliable and targeted—a kind of psychological version of "precision medicine." The upshot should be the start of a slow but important process eroding the old distinctions between psychiatry, neurology, and computational neuroscience, and at last embracing the fundamental unity of mind, body, and world.

The shape of that unity is now clear. There is a fundamental drive, instantiated by the brain, to minimize errors in our own sensory predictions. That same drive guides, and is guided by, our own internal bodily states and by a rich array of physical actions, many of them designed to gather information and reduce uncertainty. Brain structure and neurochemistry, the physiological body, our own actions, history, and practices, and the environmental settings in which we live and work, all combine and cooperate to manage the flow of prediction. Thus understood, human minds are not elusive, ghostly inner things. They are seething, swirling oceans of prediction, continuously orchestrated by brain, body, and world. We should be careful what kinds of material, digital, and social worlds we build, because in building those worlds we are building our own minds too.

Appendix:

Some Nuts and Bolts

THIS BOOK is about what predictive brains do. It's about how they support human experience, and what that implies for mind and society. To keep the narrative flowing (and because it's already been done in some detail elsewhere) I've kept technical details to a minimum. This short Appendix fills in a few of the more important gaps and offers some pointers for anyone wishing to delve deeper. Readers who are already familiar with work in this area, or simply less interested in the nuts and bolts, can safely skip this section.

Predictive brains are built of four core elements. The first is a "generative model." The second are the moment-by-moment predictions that it issues. The third are the "prediction errors" about which we have heard so much—these arise whenever incorrect or incomplete predictions attempt to meet and account for sensory evidence. The fourth are the estimations of "precision" that alter the relative impact of both sensory stimulations and predictions.

Prior knowledge, predictions, prediction errors, and precisions may sound like a lot of moving parts, but actually they are a remarkably austere toolkit from which to build everything that matters about the human mind.

Let's visit each in turn, before rounding things off with a brief look at action.

Element One: A Generative Model

The predictions that meet incoming sensory evidence are issued by a "generative model" realized by the brain. A generative model is simply a resource capable of generating (as the name implies) new instances of some kind of data, image, or structure using what it knows about how various features and properties fit together.

For example, anyone reading this book must know a language. To know and speak a language is to command a generative model for that language, since it implies the ability to build and understand brand-new never-before-seen sentences based on the general patterns that characterize that language. Similarly, a child who knows how to use a Lego kit commands something like a generative model for building brick-based structures, enabling them to build all kinds of novel structures.

Generative models can be learned by artificial neural networks trained on a huge number of real instances of the kind of data in question. This enables them to learn about the general patterns and regularities that characterize that particular data set. Once such training is complete, they become able to generate novel versions of the data—plausible "fake" instances that they have never actually been exposed to during training. To appreciate the power of a good generative model of this kind, look at the faces shown in Fig. A.1. These are two of the famous "fake celebrities" output by a generative neural network architecture that was trained for twenty days on thirty thousand images of real celebrities.

The generated faces were not simple tweaks of the kind that might be formed by adding a bigger nose, or differently colored hair to images seen in the data set used for training. Instead, the network learned a kind of probabilistic "celebrity image grammar" enabling it to construct for itself plausible novel images of celebrities who don't really exist. The

Fig. A.1 Two of the AI-generated fake celebrities

multilevel artificial neural network learned to do this by being exposed to lots of training images and trying (again and again) to generate similar images.

The use of celebrities here is something of a diversion, as the real point was to find a way to artificially generate extremely high-resolution novel images for desired categories. As it happens, online celebrity photos provided a good source of high-resolution images to work with in training. To generate novel versions of such images requires identifying and then "recomposing" key underlying patterns in some original data set. In the case of the celebrities, these are patterns in the hugely variegated space of human faces, perhaps somewhat skewed (because the training set are celebrities) toward winning smiles and symmetric features. But the same methods have been applied to all manner of data sets, ranging from potted plants to bicycles.

Generative models here arise (just as they do in predictive processing) by means of a network's ongoing attempts to generate the kinds of patterns encountered in the training data. But in these specific examples that training involved so-called adversarial methods. That means that one network (known as the generator network) tries to come up with plausible fake images while another network (known as the adversarial network) tries to catch it out—to determine whether it is a real image or a deep fake. This is a powerful method. A common comparison is between police and counterfeiters, where each side improves to keep up with the other, so that the counterfeits become very good in the end. But it is extremely unlikely that human brains use adversarial methods of this kind. Instead, they must use observation-action sequences—attempts to put sensory information to use in guiding action, that deliver multiple forms of prediction error signals whenever those attempts fail. Moreover, this learning must be capable of going on all the time, not restricted to a small window of early learning. What matters for our purposes is just that the generative model—however installed—is a learnable resource that will enable a system to self-generate plausible new versions of the kinds of data seen in training.

According to predictive processing, mature human brains encode a rich generative model of the human world. Brains like that are fully capable of generating mock versions of the sensory data for themselves. This is, in one sense, unsurprising. Reflection on dreams, vivid imaginings, and hallucinations already suggests something of this kind. There is a suggestive duality here such that to perceive the world (in this way) is to be able to imagine that world too—it is to be able to generate, using our inner resources alone, the kinds of neural response that would ensue were we in the presence of those states of affairs in the world.

Elements Two and Three: Predictions and Prediction Errors

Armed with a generative model, the brain can make informed guesses (predictions) that can be compared with the incoming signal. When it all matches up, we perceive (and understand) our worlds. But when mismatch occurs, prediction errors result. These carry information about any residual differences and enable the system to seek out a better guess—or else to act on the world, altering the inputs to fit the predictions.

In perception, the brain responds to prediction error by seeking out a better guess at what is out there in the world. Thus, suppose you are equipped with a generative model not of faces or the visual appearances of objects this time but of written English. Upon encountering stimuli like the ones shown in Fig. A.2, the predictive brain must seek out the best way to generate that fragment of sensory evidence using prior information (what the generative model knows about the

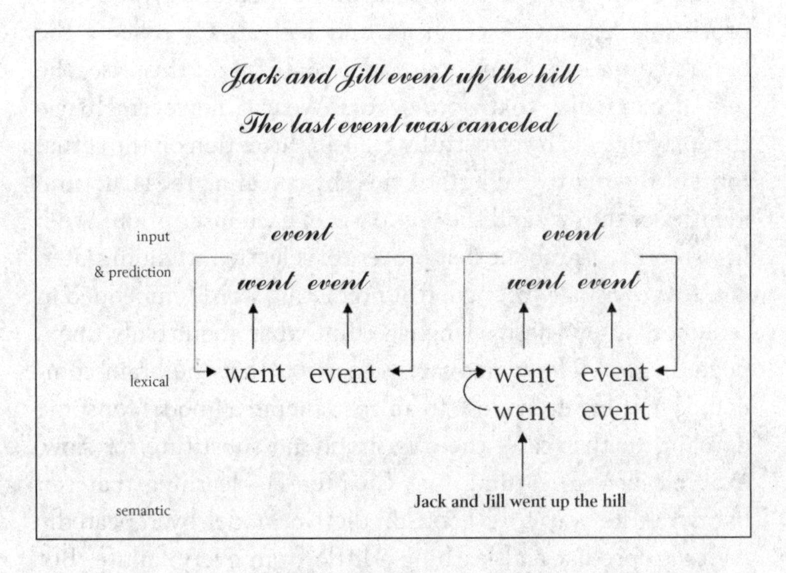

Fig. A.2 Schematic illustrating the role of priors. (Left) The word "event" is selected as the most likely cause of the visual input. (Right) The word "went" is selected as the most likely word that is (1) a reasonable explanation for the sensory input and (2) conforms to prior expectations based on semantic context.

world). The most relevant aspect of that knowledge, in this new case, is information and patterns pertaining to the forms, structures, and meanings of English.

Now look more closely at the words "went" and "event" as they occur in the top and bottom sentences. They are structurally identical, yet we effortlessly read the top one as "went" and the lower one as "event." This is because the brain's best overall guess, after a rapid flurry of prediction and error cycles (more on which shortly) is that "went" is the most likely candidate for that word slot in the top sentence while "event" is the better bet for the bottom.

This top-down influence is so potent that the structural sameness of the two inscriptions (the italicized "went" and "event") can itself be hard to spot. According to predictive processing what we see, hear, and feel always reflects the brain's best guess given some wider context—in this case, the sentence in which that word occurs. We can, however, change that playing field by effortfully focusing attention on the actual form of the individual letters, thereby revealing the structural identity of the "w" and the "ev" parts of each inscription. We'll have more to say about that process of selective attending later.

We are usually very good at predicting words embedded in sentences in our native language. But what about truly unexpected cases? The worst-case scenario is that the brain commands nothing even close to an apt generative model for some domain. In that case, there is simply no substitute for slow, example-driven learning (see Chapter 1)—learning that can bootstrap its way to a good predictive model by repeatedly trying to predict and learning a little from every failure. But in other cases, what we normally call "unexpected" is simply stuff we did not see coming at this particular moment, rather than stuff we have no idea about at all.

Suppose you are (consensually) blindfolded and gently put in a car by someone you trust. Half an hour later the blindfold is removed, and you see where you are. At the first instant the blindfold is removed, none of your brain's attempts at predicting the sensory input work. Prediction error signals follow. But thankfully you don't have to wait for slow long-term learning to install a whole new set of predictions to solve the puzzle. Instead, you blink once or twice, look around, and see your surroundings. As a birthday surprise you have been taken to a newly opened spa in a forest on the edge of town, where champagne and truffles await. How did flurries of prediction error ever manage to lead you to such a rapid understanding of an unexpected situation?

It is worth stressing that what we have imagined is actually a quite unusual case. In the normal run of things, we awake in more-or-less-familiar rooms and settings (such as being at home or at a conference hotel) and then move through a sequence of fairly predictable scene changes. For most of our lives, we harvest something very close to the streams of sensory stimulation that our brains expect. Life is, by and large, a rather predictable affair.

But we are not lost when the unexpected occurs. So, what happens when that blindfold is removed? At first, very general features of the sensory data are detected, and these are subjected to the prediction error minimizing process. That may be enough to help set the larger context, distinguishing between (for example) a cityscape, lakeside, or open sea setting. Once the gist of the general context is identified, new streams of prediction attempt to fill in further details. Prediction errors ensue until enough details emerge. All this happens extremely fast, so there is no conscious awareness of the complex give-and-take between predictions, errors, and revised predictions.

This means we perceive the "woods before the trees" using coarse, rapidly processed cues to identify larger patterns and contexts that then narrow down the search. As this process unfolds, increasingly specific predictions result in increasingly specific kinds of error, and those errors help select new predictions designed to quash them. In this way rich and meaningful experience is pressed from the dance between predictions and prediction errors.

Action, in these frameworks, is constructed in much the same way as perception. In each case, the goal is to minimize errors in the prediction of the sensory stream. But in constructing action, it responds to a subset of those errors (proprioceptive prediction errors) by moving the body to make those predictions come true. Proprioception, as we saw in the main text, is just a technical name for the sensory networks that allow us to sense our own movements and the locations of our bodily parts in space. According to predictive processing (active inference), movement occurs when the brain predicts a pattern of proprioceptive signals that are not yet actual, and then eliminates the resulting proprioceptive prediction errors by systematically bringing those movements about. Predictions of the sensory consequences of an intended action thus play the role more normally assigned to motor instructions—the role of bringing about the actions that we want to perform. In this way action involves a kind of self-fulfilling prophecy.

Element Four: Precisions

We have seen how perception and action co-emerge from an ongoing dance between predictions and prediction errors. That dance is orchestrated by our final key player—the precisions that weight predictions and prediction errors, altering neural and bodily responses accordingly.

Take an even closer look at the wordforms in Fig. A.2, this time attending to the way the letters "EV" occur in the inscription of "event," and the way the letter "W" occurs in the inscription of "went." You may notice that seen as an occurrence of "W" in "went," the letter in the top ("input") sentence is somewhat deformed, while the occurrence of "EV" in "event" is not. This is not obvious at first glance. This is because the good overall fit of the best top-level sentence model leads the brain to treat the slight deformity of the word "went" as unimportant noise, which it promptly ignores. Just as in the hallucination of "White Christmas" in Chapter 1, strong expectations (strong priors, enshrined in the generative model) sometimes allow the brain actively to carve out an expected form or signal from inadequate or even somewhat contrary sensory evidence.

Yet closer inspection clearly revealed the letter-level deformity. So what changed when we attended more closely to the form of the letters? The answer lies in the brain's remarkable ability to alter the relative weightings on key elements of its own predictive regime. By increasing or decreasing these "precision-weightings," the impact of certain predictions or of certain bits of sensory evidence can be amplified or dampened. This allows the brain to "turn up the volume" on information that is estimated as both important (task-salient) and reliable. What we informally think of as "attention" is implemented in these systems by mechanisms that alter these precision-weightings. By actively attending to the letter form, we increased the precision-weighting on that specific fragment of visual evidence, and (thereby, as this is a zero-sum game) somewhat decreased the effective weighting on the higher-level predictions that might otherwise trump whatever fails to conform to the overall context. Attention tends in this way to reverse some of the effects of top-down prediction and expectation. That's how the brain allows the letter form visual evidence to speak to us a little bit more strongly—enabling us to detect even slight deformities that we would otherwise miss.

More generally still, attention can dampen or amplify select aspects of the neuronal guessing game, tuning patterns of response in ways that better serve some specific task or goal.

I recently had the experience of searching for my car keys on a crowded work surface. No sign of them at all. Yet I was quite sure I had laid them down there just seconds ago. Then I recalled that I had recently changed the key fob—the fob that used to be yellow is now bright pink (a souvenir from a Pet Shop Boys tour gig). The keys, previously invisible, immediately jumped out at me.

The mechanism here has been crucial to our story. My revised prediction that the fob should be pink altered the precision-weightings on specific kinds of sensory information, allowing me to search the scene in a new way—a search in which the specific shade of Pet Shop Boys pink was rendered especially salient. This enabled my brain to resolve uncertainty of a very specific kind—uncertainty about the location of anything of that shade on the work surface. The pink prediction altered the response thresholds of key neuronal populations in the visual processing stream, thereby reconfiguring the search and enabling me to find the keys.

Variable precision-weighting is the single most powerful tool in the predictive processing toolkit. To begin to get an intuitive sense of that remarkable power, we can compare it to the humble headlamp. The first time I strapped one on I was amazed. Where moments earlier I had been scrambling to find my way around during an unexpected power cut, I was suddenly able to see pretty much whatever I wanted, with no conscious thought or effort. The experience was quite unlike that of aiming a flashlight. Instead of having two things to do (aim the flashlight, and then act on the world) I was just acting on the world. It felt like the world was revealing itself to me, but (and this was the almost miraculous bit) in just the right ways and at just the right times needed to support my current actions and purposes.

The analogy is imperfect, but precision-weighting shares something of this profile. In the brain, precision-weighting alters patterns of post-synaptic influence (the strength of the signals passed on after the synapse "fires"). This means that specific signals can be selected for enhanced impact. The signals selected for this special treatment will be ones that are expected to be both reliable and important for the task at hand. A good way to think of precision is as indicating something like "estimated value of this information, in this context, for this task." Precision variations of this kind underpin both conscious and automatic deployments of attention. As we go about our daily lives, precision is estimated again and again, across multiple brain areas, and at different levels of processing. In this way, variable precision-weighting provides a flexible means for the brain to adjust its responses according to task and context, rendering these architectures quite astonishingly flexible and fluid.

It is worth noting that there is nothing mysterious about the learning process behind variable precision-weighting. The brain learns how and when to vary its precision estimations as part and parcel of the process by which it acquires the generative models themselves. Those models and associated estimations of precision are learned by repeated exposure to flows of sensory information in the context of trying to act in the world. By trying to predict those flows as we seek to act, our brains learn to extract helpful patterns of many kinds and at many levels. But when precision assignments go wrong, our brains become confident of the wrong things, and lose confidence in the right things. This is the pernicious kind of misfiring that sometimes results in mistakes (such as functional pain and paralysis—see Chapter 2) and mental illness.

Such mistakes and misfirings, and our tendency to sometimes see only what we expect to see, are the darker side of the predictive brain in action. But the benefits are legion. We can discern subtle patterns through seas of noise and ambiguity, and we can reconfigure our own processing routines in ways that reflect current tasks and context. The whole embodied organism is continuously organizing its responses around precision-weighted prediction error. The attempt to minimize those errors then drives both perception and action, locking them into a tight and mutually empowering embrace.

Such, then, is the core toolkit that the brain uses to orchestrate our embodied interactions with the world. As the complexity and depth of the guiding predictive model increases, that same toolkit (see Chapter 6) supports reasoning about multiple futures—ways things will alter if (but only if) we select one action rather than another. In that way, creatures like us become truly enmeshed in the worlds we live in, selecting actions that make use of tools, practices, and environmental opportunities to reduce future error, bringing us closer to our goals. It is not merely our predictive brains but this whole complex dance that makes us who and what we are.

Acknowledgments

This book ranges over many territories, from Philosophy and Cognitive Science (my home turf) to Psychology, Neuroscience, Machine Learning, Medicine, Psychiatry, and beyond. Such extensive landscapes are challenging to negotiate. I've tried to guard against outright errors by getting experts in each field to cast an eye over key parts and passages, and in a few cases over entire drafts. For such extreme labors a huge thank-you goes out to Anil Seth, Rob Clowes, and Jakob Hohwy. Jon Stone (Professor of Neurology at the University of Edinburgh) offered generous and helpful comments that have greatly improved the sections relating to functional disorders. For errors that remain, I apologize. I've also added copious notes and references so the reader can explore further and come to their own conclusions.

My greatest intellectual debt is, as always, to Daniel Dennett, whose ideas, friendship, and support have been a constant in my life. The ideas and themes pursued here also owe much to my new (and old) friends and colleagues at the University of Sussex, U.K., especially to Anil Seth and to Sarah Garfinkel (now at University College London); to my old crew from the University of Edinburgh, especially Peggy Series and Robert McIntosh; to my present and past colleagues at Macquarie University in Australia, especially Richard Menary and John Sutton; to the extended network working on the front lines of predictive processing, especially Karl Friston, Jakob Hohwy, Lucia Melloni, Micah Allen, Giovanni Pezzulo, Julian Kiverstein, Axel Constant, Erik Rietveld, and Lisa Feldman Barrett; and to the amazing X-Team, Mark Miller, David Carmel, Sam Wilkinson, Frank Schumann, George Deane, and Kate Nave, who were my collaborators on the Horizon 2020 European Union ERC Advanced Grant *Expecting Ourselves: Embodied Prediction and the Construction of Conscious Experience* (XSPECT—DLV-692739). Many of the ideas presented here took shape while I was Principal Investigator on that four-year project. You can find out more about that work at www.x-spect.org.

My personal debts are far too many to count. But a special mention for my brothers Gordon and David Clark, my brother Jimmy who died far too soon, and all their families. For my partner Alexa Morcom, her amazing parents Caroline and John Morcom, and all her family. For my dear friends Gill Banks, Mark Sayers, Rachael and Colin Mackenzie, Nigel Davies, Ian Davies, and Eric Braund.

Thanks also to the University of Edinburgh and the University of Sussex for allowing me precious time to work on this and other projects. Sincere thanks to my helpful and diligent copy editor, Fred Chase, and to my editors Edward Kastenmeier, Laura Stickney, and Rowan Cope. Your comments and suggestions have been invaluable. This book is much better as a result. Thanks also to Sam Fulton and Chris Howard-Woods for your amazing help in the final stages of production, and (for the audio book recordings) to Paul James, Jack Levy, and all at Pier Productions, Brighton, United Kingdom.

Notes

PREFACE: SHAPING EXPERIENCE

xii A 2012 study found: The medical intern study is reported in Rothberg, M. B., et al., "Phantom Vibration Syndrome Among Medical Staff: A Cross Sectional Survey," *BMJ (British Medical Journal)* 341 (December 15, 2010). The work with college undergraduates was by Drouin, M., Kaiser, D. H., and Miller, D. A., "Phantom Vibrations Among Undergraduates: Prevalence and Associated Psychological Characteristics," *Computers in Human Behavior* 28(4) (2012): 1490–1496. See also Sauer, V. J., et al., "The Phantom in My Pocket: Determinants of Phantom Phone Sensations," *Mobile Media & Communication* 3(3) (2015): 293–316.

1. UNBOXING THE PREDICTION MACHINE

4 It has long been known that hallucinations: An early demonstration can be found in Ellison, D. G., "Hallucinations Produced by Sensory Conditioning," *Journal of Experimental Psychology* 28 (1941): 1–20. By repeatedly pairing seeing a bulb dimly light up with hearing a simple tone, experimenters induced subsequent hallucinatory experiences. Once they were conditioned by the training, hearing the tone led to subjects reporting spotting a very faint light.

7 This was probably because: The Marr work was neatly laid out in his landmark book *Vision: A Computational Investigation into the Human Representation and Processing of Visual Information* (Cambridge: MIT Press, 1982).

8 Even into the twenty-first century: See, for example, Riesenhuber, M., and Poggio, T., "Models of Object Recognition," *Nature Neuroscience* 3 (Suppl) (2000): 1199–1204. For an important early version of the feedforward view, see Hubel, D. H., and Wiesel, T. N., "Receptive Fields and Functional Architecture in Two Nonstriate Visual Areas (18 and 19) of the Cat," *Journal of Neurophysiology* 28 (1965): 229–289. For some of the first conclusive empirical tests that began to show the inadequacy of this feedforward picture, see Egner, T., Monti, J. M., and Summerfield, C., "Expectation and Surprise Determine Neural Population Responses in the Ventral Visual Stream," *Journal of Neuroscience* 30(49) (2010): 16601–16608. See also

Petro, L., Vizioli, L., and Muckli, L., "Contributions of Cortical Feedback to Sensory Processing in Primary Visual Cortex," *Frontiers in Psychology* 5 (2014): 1223.

8 The number of neuronal connections: For example, after light enters the eye, signals are first passed to the lateral geniculate nucleus (LGN), which then routes information onward to an area known as V1. But this forward-flowing stream is just a small part of the story. The bulk (perhaps 80 percent) of the input to LGN actually comes from elsewhere in the brain, with much of it feeding back downward from V1. For these connectivity estimates, see Budd, J. M. L., "Extrastriate Feedback to Primary Visual Cortex in Primates: A Quantitative Analysis of Connectivity," *Proceedings of the Royal Society B: Biological Sciences* 265 (1998): 1037–1044. See also Raichle, M. E., and Mintun, M. A., "Brain Work and Brain Imaging," *Annual Review of Neuroscience* 29 (2006): 449–476.

8 This wiring runs in the opposite direction: For some excellent discussion, and a compelling experimental demonstration of predictive processing in action, see Muckli, L., et al., "Contextual Feedback to Superficial Layers of V1 Report," *Current Biology* 25 (2015): 2690–2695.

8 The brain, weighing in at about 2 percent: The energy consumption figures are from Raichle, M., "The Brain's Dark Energy," *Science* 314 (2006): 1249–1250.

9 artificial intelligence pioneer Patrick Winston: The claim that we were "clueless" about the true role of the recurrent architecture is made in Winston, P., "The Next 50 Years: A Personal View," *Biologically Inspired Cognitive Architectures* 1 (2012): 92–99. We were not, however, quite as clueless as Winston suggested even then, as good ideas about the true role of all that downward wiring had been around for quite a while. For some examples, see references in the following note.

10 the last ten to fifteen years: A key publication within that lineage was Friston, K., "A Theory of Cortical Responses," *Philosophical Transactions of the Royal Society of London B Biological Sciences 29,* 360(1456) (2005): 815–836. Earlier treatments include Mumford, D., "On the Computational Architecture of the Neocortex II: The Role of Cortico-Cortical Loop," *Biological Cybernetics* 66 (1992): 241–251; and Lee, T. S., and Mumford, D., "Hierarchical Bayesian Inference in the Visual Cortex," *Journal of Optical Society of America A,* 20(7) (2003): 1434–1448. See also Hinton, G. E., et al., "The Wake-Sleep Algorithm for Unsupervised Neural Networks," *Science* 268 (1995): 1158–1160.

10 Despite its intuitive appeal: The caveat is important. Much of the knowledge that grounds predictions comes from a lifetime of learning. But some will have been pre-installed by evolution as part of the brain's basic structure and connectivity patterns. See Teufel, C., and Fletcher, P. C., "Forms of Prediction in the Nervous System," *Nature Reviews Neuroscience* 21(4) (2020): 231–242. Completing that more nuanced picture, some of the knowledge acquired during lifetime learning may subsequently become compressed (in a process known as "amortized inference") into fast, efficient links that sacrifice flexibility for speed. These fast, frozen linkages may (for example) help us to rapidly appreciate the gist of a scene, paving the way for the com-

plex give-and-take of prediction and prediction error signaling that will be our primary focus. For more on amortized inference, see Tschantz, A., et al., "Hybrid Predictive Coding: Inferring, Fast and Slow," *arXiv*:2204:02169v2 (2022). For an application to planning, see Fountas, Z., et al., "Deep Active Inference Agents Using Monte-Carlo Methods," *Advances in Neural Information Processing Systems 33* (eds.) Larochelle, H., et al. (Red Hook, N.Y.: Curran Associates, 2020), pp. 11662–11675.

11 He was also interested in theories: See Hermann von Helmholtz, *Handbuch der physiologischen optic*, in J. P. C. Southall (ed.), (English trans.), Vol. 3 (New York: Dover, 1860/1962).

13 "controlled hallucination": It's unclear exactly who originated this striking phrase. It may have been the machine learning pioneer Max Clowes.

14 Linear predictive coding: See Shannon, C., "A Mathematical Theory of Communication," *Bell System Technical Journal* 27 (July, October, 1948): 379–423, 623–656.

14 Telecommunications research: One especially prescient version emerged in a pair of early papers by Peter Elias of the Institute of Radio Engineers. These 1955 papers were rediscovered in 1965 by Manfred Schroeder and Bishnu Atal, again working at Bell Laboratories. See Atal, B. S., "The History of Linear Prediction," *IEEE Signal Processing Magazine* 161 (2006): 154–161.

15 In 1959, the world: See Musmann, H., "Predictive Image Coding," in *Image Transmission Techniques* (Advances in Electronics and Electron Physics, Suppl 12), ed. W. K. Pratt (New York: Academic Press, 1979), 73–112.

16 Human brains seem to benefit: See, for example, Friston, K., "The Free-Energy Principle: A Rough Guide to the Brain?," *Trends in Cognitive Sciences* 13(7) (July 2009): 293–301. For some of the more detailed work linking this broad picture to the circuitry of the human brain, see Bastos, A. M., et al., "Canonical Microcircuits for Predictive Coding," *Neuron* 76(4) (November 2012): 695–711. For a fairly balanced view of the state of the neural evidence, see Walsh, K. S., et al., "Evaluating the Neurophysiological Evidence for Predictive Processing as a Model of Perception," *Annals of the New York Academy of Sciences* 1464(1): 2020: 242–268. See also De Lange, F., Heilbron, M., and Kok, P., "How Do Expectations Shape Perception?," *Trends in Cognitive Sciences*, 22(9) (June 2018): 764–779.

18 In the case at hand, masked presentations: See Biderman, D., Shir, Y., and Mudrik, L. B., "Unconscious Top-Down Contextual Effects at the Categorical but Not the Lexical Level," *Psychological Science* 31(6) (2020): 663–677.

18 When lit from behind and viewed: There are good examples of this effect viewable online—see, for example, the video at www.richardgregory.org/experiments/video/chaplin.htm. For more on the Hollow Mask illusion, see Gregory, R. L., "Knowledge in Perception and Illusion," *Philosophical Transactions of the Royal Society London, B* 352 (1997): 1121–1128.

19 This so-called Mooney image: Mooney images are named after the psychologist Craig Mooney, who, in 1957, handcrafted a number of such stimuli as simple tools to investigate the use of minimal information to create a

meaningful visual perception. The work appeared in Mooney, C. M., "Age in the Development of Closure Ability in Children," *Canadian Journal of Psychology* 11(4) (1957): 219–226. The image used in the text is from Rubin, N., Nakayama, K., and Shapley, R., "The role of Insight in Perceptual Learning: Evidence from Illusory Contour Perception," *Perceptual Learning*, Fahle, M., and Poggio, T. (eds.) (Cambridge: MIT Press, 2002).

20 The picture looks different the second time: We can say a bit more about this. Once you have seen the original (non-Mooney) image, then when you again see the Mooney version, you benefit from an enriched knowledge base that guides the way you visually explore the image. Crucially, the responses of neurons that detect kitten-relevant local features in early processing are also sharpened. The incoming sensory information is unchanged (it is still the same Mooney image) but that information is now well predicted by a much higher-level body of knowledge, and the responses of early feature detectors are altered accordingly. For more on the way prediction decodes Mooney images, see Teufel, C., Dakin, S. C., and Fletcher, P. C., "Prior Object-Knowledge Sharpens Properties of Early Visual Feature-Detectors," *Nature Scientific Reports* 8, 10853 (2018).

20 It was invented back in the early 1970s: For the original paper on sine-wave speech, see Remez, R. E., et al., "Speech Perception Without Traditional Speech Cues," *Science* 212 (1981): 947–950. A contemporary take on the phenomenon, with sound file examples, can be found in "An Introduction to Sine-Wave Speech" by Matt Davis, available online.

23 The experiment was successfully replicated: The song choice reflects the earliest origins of this work, which date back to 1964 when Bing Crosby was still among the world's best-known crooners. The 1964 paper was Barber, T. X., and Calverey, D. S., "An Experimental Study of Hypnotic (Auditory and Visual) Hallucinations," *Journal of Abnormal and Social Psychology* 68 (1964): 13–20. There, the effect was described as a "hypnotic hallucination." The more recent experiment with undergraduates is reported in Merkelbach, H., and van de Ven, V., "Another White Christmas: Fantasy Proneness and Reports of 'Hallucinatory Experiences' in Undergraduate Controls," *Journal of Behavior Therapy and Experimental Psychiatry* 32 (2001): 137–144. The larger follow-up study was van de Ven, V., and Merkelbach, H., "The Role of Schizotypy, Mental Imagery, and Fantasy Proneness in Hallucinatory Reports of Undergraduate Students," *Personality and Individual Differences* 35 (2003): 889–896.

23 There is also some evidence that both false song detection: Stress and caffeine enhance the tendency to hallucinate the onset of "White Christmas." See Crowe, S. F., et al., "The Effect of Caffeine and Stress on Auditory Hallucinations in a Non-Clinical Sample," *Personality and Individual Differences* 50 (2011): 626–630. The work on schizophrenia is reported in Mintz, S., and Alpert, M., "Imagery Vividness, Reality Testing and Schizophrenic Hallucinations," *Journal of Abnormal Psychology* 79 (1972): 310–316, and followed up in Young, H. F., et al., "The Role of Brief Instructions and Suggestibility in the Elicitation of Auditory and Visual Hallucinations in Normal and Psychiatric Subjects," *Journal of Nervous and Mental Disease* 175 (1987): 41–48.

23 What seems indisputable: Hearing "White Christmas" in these circumstances is not in itself a sign of psychosis or abnormality, so much as a reflection of the way our brains construct ordinary daily experience. Instead, work on the predictive brain provides a strong hint that typical and atypical forms of human experience are constructed in remarkably similar ways, an insight at the heart of promising new work in "computational psychiatry." More on this in Chapter 2.

25 Brains that make different assumptions: For a detailed analysis, see Witzel, C., Racey, C., and O'Regan, J., "Perceived Colors of the Color-Switching Dress Depend on Implicit Assumptions About the Illumination," *Journal of Vision* 16(12) (2016): 223.

26 The authors of the study conjecture: The work on owls and larks can be found in Wallisch, P., "Illumination Assumptions Account for Individual Differences in the Perceptual Interpretation of a Profoundly Ambiguous Stimulus in the Color Domain: 'The Dress,'" *Journal of Vision* 17(4) (2017): 1–14.

28 As those attempts continue: In the visual domain, an important proof of principle for prediction-driven learning was an artificial neural network built by computational neuroscientists Rajesh Rao and Dana Ballard in the final years of the twentieth century. This network was fed large numbers of sample images drawn from pictures of natural scenes—including pictures of zebras, swans, monkeys, and forests. The image samples were fed to a simple prediction architecture in which one level was busy trying to predict the current activity at the level below. Over time, the network (which started out knowing nothing at all) learned about patterns in the natural images, proving that the attempt to predict can install the knowledge needed to succeed at predicting. See Rao, R., and Ballard, D., "Predictive Coding in the Visual Cortex: A Functional Interpretation of Some Extra-Classical Receptive-Field Effects," *Nature Neuroscience* 2(1) (1999): 79.

31 Fig. 1.5 On the traditional view: The illustration is inspired by a user-friendly online introduction to predictive processing by Curtis Kelly. Kelly works with the Japan Association of Language Teaching (JALT) Mind, Brain, and Education special interest group. The picture is from a primer on predictive processing published in *Bulletin of the JALT Mind, Brain, and Education SIG* 6(10), (October 1, 2020). The full primer is available at https://www.mindbrained.org/october-2020-predictive-processing/.

32 Brains are prediction machines: Other books targeting the core picture of brains as prediction machines include the excellent general introduction by Jakob Hohwy, *The Predictive Mind* (New York: Oxford University Press, 2013); and my own *Surfing Uncertainty: Prediction, Action, and the Embodied Mind* (New York: Oxford University Press, 2016). The view from cognitive neuroscience is elegantly captured in Anil Seth, *Being You: A Science of Consciousness* (Penguin, UK, 2021). A wonderful treatment highlighting the role of bodily prediction is Lisa Feldman-Barrett, *How Emotions Are Made: The Secret Life of the Brain* (New York: Houghton Mifflin Harcourt, 2018).

2. PSYCHIATRY AND NEUROLOGY: CLOSING THE GAP

33 A report in the *British Medical Journal:* The case of the construction worker with the nail through his boot was first shared with me by Mick Thacker, a leading U.K. pain researcher. It is reported in Fisher, J. P., Hassan, D.T., and O'Connor N. M., *British Medical Journal* 310 (1995): 70. See also Dimsdale, J. E., and Dantzer, R. A., "Biological Substrate for Somatoform Disorders: Importance of Pathophysiology," *Psychosomatic Medicine* 69(9) (2007): 850–854.

35 Predictive processing provides: Predictive processing here names the class of computational models of the predictive brain with which this book is mostly concerned. That class of models, as remarked earlier in the text, is also known as "active inference"—so named because they depict perception as inference and unify the treatment of perception and action in a distinctive way. That new unity is the topic of Chapter 3. It is worth noticing that the phrase "predictive processing" is sometimes used in an even wider sense, to name any computational story that treats the brain as a prediction machine.

36 In the U.K. alone, a 2016 meta-analysis: See Fayaz, A., et al., "Prevalence of Chronic Pain in the UK: A Systematic Review and Meta-Analysis of Population Studies," *BMJ Open* 6 (2016): e010364.

37 In 2016 a third category was added: For all these definitions, and some of the recent history, see Cohen, S. P., Vase, L., and Hooten, W. M., "Chronic Pain: An Update on Burden, Best Practices, and New Advances," *Lancet* 397(10289) (2021): 2082–2097. For a handy historical overview of the development of our conceptions of pain, see Raffaeli, W., and Arnaudo, E., "Pain as a Disease: An Overview," *Journal of Pain Research 10* (2017).

37 A repeated theme: For helpful introductions to the contemporary neuroscience of pain, see N. Twilley's *New Yorker* magazine (Annals of Medicine) piece "The Neuroscience of Pain" (July 2018); and Y. Bhattacharjee's "A World of Pain," *National Geographic* (January 2020). I'd also recommend leading pain researcher professor Irene Tracey's short introduction to key issues, "Finding the Hurt in Pain," in *Cerebrum: The Dana Forum on Brain Science* (December 2016). For a review of some key experiments, see Atlas, L. Y., and Wager, T. D., "How Expectations Shape Pain," *Neuroscience Letters* 520 (2012): 140–148. For a very thorough treatment using the resources of predictive processing, see Kiverstein, J., Kirchhoff, M. D., and Thacker, M., "An Embodied Predictive Processing Theory of Pain Experience," *Review of Philosophy and Psychology* (2022).

37 In work dating back to the 1990s: For a recent review, see Denk, F., McMahon, S. B., and Tracey, I., "Pain Vulnerability: A Neurobiological Perspective," *Nature Neuroscience* 17 (2014): 192.

37 In one striking fMRI study: The study involving religious imagery appears as Wiech, K., et al., "An fMRI Study Measuring Analgesia Enhanced by Religion as a Belief System," *Pain* 139(2) (2009): 467–476.

38 In one such study, experimenters used heat stimuli: See Brown, C. A., et al., "Modulation of Pain Ratings By Expectation and Uncertainty: Behavioral Characteristics and Anticipatory Neural Correlates," *Pain* 135

(2008): 240–250. See also F. Fardo, et al., "Expectation Violation and Attention to Pain Jointly Modulate Neural Gain in Somatosensory Cortex," *Neuroimage* 153 (2017): 109–121.

39 In one experiment, different subliminally presented: See Jensen, K., et al., "Classical Conditioning of Analgesic and Hyperanalgesic Pain Responses Without Conscious Awareness," *Proceedings of the National Academy of Sciences of the United States of America* 112(25) (2015): 7863–7867.

40 Placebo-induced changes have been shown: Studies showing the spinal cord impacts of placebos include Eippert, F., et al., "Direct Evidence for Spinal Cord Involvement in Placebo Analgesia," *Science* 326 (2009): 404; and Geuter, S., and Buchel, C., "Facilitation of Pain in the Human Spinal Cord by Nocebo Treatment," *Journal of Neuroscience* 33 (2013): 13784–13790.

40 A striking example is "placebo analgesia": An excellent review of placebo research can be found in Büchel, C., et al., "Placebo Analgesia: A Predictive Coding Perspective," *Neuron* 81(6), (2014): 1223–1239.

41 In a laboratory setting, some forms of hypnosis: See Facco, E., "Hypnosis and Anesthesia: Back to the Future," *Minerva Anestesiologica* 82(12) (2016): 1343–1356. For the "dental pulp" experiments, see Facco, E., et al., "Effects of Hypnotic Focused Analgesia on Dental Pain Threshold," *International Journal of Clinical Experimental Hypnosis* 59 (2011): 454–468.

41 A large 2013 survey: See https://www.nhs.uk/news/medical-practice/survey-finds-97-of-gps-prescribe-placebos/.

41 For even if only a very small fraction: See Colloca, L., and Miller, F. G., "The Nocebo Effect and Its Relevance for Clinical Practice," *Psychosomatic Medicine* 73(7), (2011): 598–603. For an interesting exploration of some ways to combat the problem of inducing nocebo effects, see Nestoriuc, Y., et al., "Informing About the Nocebo Effect Affects Patients' Need for Information About Antidepressants—An Experimental Online Study," *Frontiers in Psychiatry* 12 (2021).

42 In one important recent study, experimenters showed: The study showing self-reinforcing pain expectations is Jepma, M., et al., "Behavioural and Neural Evidence for Self-Reinforcing Expectancy Effects on Pain," *Nature Human Behaviour* 838(2) (2018): 838–855.

42 In the studies, participants were first shown arbitrary abstract visual cues: The high readings were shown as 73–93 percent of the thermometer scale, while the low ones ranged from 25–51 percent of the scale.

42 By keeping the actual intensity of the heat: In case you are worrying about the ethics of inducing pain even in paid human subjects, it's notable that heat pain receptors in the skin activate at around 38–42 C (100–107 F). Tissue damage, by contrast, kicks in only at about 45 C (113 F). Nature has thus built a kind of prediction regime right into the basic operation of the heat nociceptors, since they respond most strongly when the actual stimulus suggests the increasing statistical likelihood of "damage in the near future"—a profile that has been nicely described as a kind of inbuilt predictive probabilistic risk assessment. See Morrison, I., Perini, I., and Dunham, J., "Facets and Mechanisms of Adaptive Pain Behavior: Predictive Regulation and Action," *Frontiers in Human Neuroscience* 7 (2013): 755.

42 Specifically, they were looking for: Specifically, they were looking for the

neurologic pain signature (NPS). This is a complex brain imaging (fMRI) signature that is claimed to be sensitive to, and specific to, physical pain. The NPS is introduced in Wager, T. D., et al., "An fMRI-Based Neurologic Signature of Physical Pain," *New England Journal of Medicine* 368 (2013): 1388–1397. The NPS has since been joined by the (complementary) stimulus intensity independent pain signature-1 (SIIPS1), which seeks—using multivariate pattern analysis—to capture the higher-level components involved in judgment and report instead. It is not clear, from a predictive processing perspective, to what extent these higher- and lower-level effects can ever be experientially distinguished or fully disentangled. But the idea that differing inner balances are at work in various cases is important. For a fascinating discussion of these attempts to find neural signatures of pain, see Woo, C., et al., "Quantifying Cerebral Contributions to Pain Beyond Nociception," *Nature Communications* 8, 14211 (2017).

43 Functional Disorders: Sincere thanks to Jon Stone, Professor of Neurology at the University of Edinburgh, for generous and helpful comments that have greatly improved the treatment of functional disorders that follows.

44 In traditional psychiatry, diagnoses are made: For a comprehensive review, see Murphy, D., "Philosophy of Psychiatry," in *The Stanford Encyclopedia of Philosophy* (Fall 2020 edition), Edward N. Zalta (ed.).

44 Arising at the crossroads of neuroscience: See for example Huys, Q., Maia, T., and Frank, M., "Computational Psychiatry as a Bridge from Neuroscience to Clinical Applications," *Nature Reviews Neuroscience* 19(3) (2016): 404; and Montague, P. R., et al., "Computational Psychiatry," *Trends in Cognitive Sciences* 16 (2012): 72–80.

45 In much of the literature, this same distinction: Thanks to Jon Stone for reminding me of the importance of making this clear. For an excellent (nicely humorous) treatment, see Stone, J., and Carson, A., "'Organic' and 'Non-organic': A Tale of Two Turnips," *Practical Neurology* 17 (2017): 417–418. Of course, some danger remains in the preferred contrast between "functional" and "structural" since functional disorders do, clearly, involve structural changes albeit of a somewhat subtle kind. Alterations to the precision-weighting regime must be realized by the brain and nervous system. But marking some kind of contrast still seems useful (perhaps one day it will not).

45 They are the second most common reason: See Stone, J., et al., "Who Is Referred to Neurology Clinics?—The Diagnoses Made in 3,781 New Patients," *Clinical Neurology and Neurosurgery* 112(9) (November 2010): 747–751. See also Carson, A., and Lehn, A., "Epidemiology," in Hallett, M., Stone, J., and Carson, A. (eds.), *Handbook of Clinical Neurology*, Vol. 139, *Functional Neurologic Disorders* (Amsterdam: Elsevier, 2016), 47–60.

45 Functional disorders can present: The list of examples of functional symptoms (which is by no means complete) is drawn from Edwards, M. J., et al., "A Bayesian Account of 'Hysteria,'" *Brain* 135 (Pt. 11) (2012): 3495–3512.

46 What could reasonably lead medical practitioners: For a review of the distinctive patterns seen in functional disorders, and a compelling argument (more on which later in this chapter) that chronic pain often shares some of the same features and etiology, see Bergh, O., et al., "Symptoms and

the Body: Taking the Inferential Leap," *Neuroscience and Biobehavioral Reviews* 74 (2017): 185–203.

48 Stone recounts the tale of a teenager: See Yeo, J. M., Carson, A., and Stone, J., "Seeing Again: Treatment of Functional Visual Loss," *Practical Neurology* 19(2) (April 2019): 168–172. The case study is also discussed by David Robson in *The Expectation Effect* (Canongate, UK, 2022).

51 This would skew the impact: Such skewing could have many physiological causes, since precision-weighting is accomplished by a variety of interacting mechanisms including the neuromodulators dopamine, serotonin, noradrenaline, and acetylcholine, each of which seems to modulate precision in different ways. For some careful work on the role and realizers of precision estimations, see Marshall, L., et al., "Pharmacological Fingerprints of Contextual Uncertainty," *PloS (Public Library of Science) Biology* 14(11) (2016): e1002575.

51 Human experience, in such cases: See Powers, A. R., Bien, C., and Corlett, P. R., "Aligning Computational Psychiatry with the Hearing Voices Movement," *JAMA Psychiatry* 75(6) (2018): 640–641.

51 Patients with these tremors: The link between looking times and functional tremor appears in van Poppelen, D., et al., "Attention to Self in Psychogenic Tremor," *Movement Disorders* 26(14), (2011): 2575–2576.

52 But if the tremor is actually: But see Matthews, J., et al., "Raised Visual Contrast Thresholds with Intact Attention and Metacognition in Functional Motor Disorder," *Cortex* 125 (2020): 161–174.

52 The doctor's request diverts attention: For more on this test (and others), see Greiner, C., Schneider, A., and Leemann, B., "Functional Neurological Disorders: A Treatment-Focused Review," *Swiss Archives of Neurology, Psychiatry and Psychotherapy* 167(8) (2016): 234–240.

53 This is a clever test: Another example is the "tremor entrainment test," in which a person with functional hand tremor is asked to copy a movement of the affected hand using the good hand. In cases of functional tremor, the relocation of attention causes the tremor in the afflicted hand to disappear. This does not occur in cases of nonfunctional tremor such as those caused by Parkinson's disease. See Stone, J., Burton, C., and Carson, A., "Recognising and Explaining Functional Neurological Disorder," *British Medical Journal* 371 (2020): m3745. See also Finkelstein, S. A., et al., "Functional Neurological Disorder in the Emergency Department," *Academic Emergency Medicine* 28(6) (June 2021): 685–696.

53 The early (1908) literature: See Hoover, C. F., "A New Sign for the Detection of Malingering and Functional Paresis of the Lower Extremities," *Journal of the American Medical Association* 51 (1908): 746–747.

54 Further support for the attentional hypothesis: See McIntosh, R. D., et al., "Attention and Sensation in Functional Motor Disorder," *Neuropsychologia* 106 (April 2017): 207–221.

55 that symptoms across a wide range of conditions: For a wonderful philosophical treatment of the varieties of pain and pain experience, see Colin Klein, *What the Body Commands: The Imperative Theory of Pain* (Cambridge: MIT Press, 2015). And for a compelling predictive processing model of acute pain, see Morrison, I., Perini, I., and Dunham, J., "Facets

and Mechanisms of Adaptive Pain Behavior: Predictive Regulation and Action," *Frontiers in Human Neuroscience* 7 (2013): 755.

55 Similar results were found: See Bergh, O., et al., "Symptoms and the Body: Taking the Inferential Leap," *Neuroscience and Biobehavioral Reviews 74* (2017): 185–203.

55 Multiple studies suggest that asthma patients: See Janssens, T., et al., "Inaccurate Perception of Asthma Symptoms: A Cognitive-Affective Framework and Implications for Asthma Treatment," *Clinical Psychology Review* 29(4): (June 2009): 317–327. See also Janssens, T., and Ritz, T., "Perceived Triggers of Asthma: Key to Symptom Perception and Management," *Clinical and Experimental Allergy: Journal of the British Society for Allergy and Clinical Immunology* 43(9) (September 2013): 1000–1008; and Teeter, J. G., and Bleecker, E. R., "Relationship Between Airway Obstruction and Respiratory Symptoms in Adult Asthmatics," *Chest* 113(2) (February 1998): 272–277.

56 "after several bouts of back pain": The quote is from a *New Scientist* (August 28, 2018) piece by Helen Thomson entitled "The Back Pain Epidemic: Why Popular Treatments Are Making It Worse."

56 In a certain sense, chronic pain: See Raffaeli, W., and Arnaudo, E., "Pain as a Disease: An Overview," *Journal of Pain Research* 10 (2017): 2003–2008.

58 Rather than weakened predictions: The picture of autism spectrum condition as involving underweighting our own predictions was suggested in Pellicano, E., and Burr, D., "When the World Becomes Too Real: A Bayesian Explanation of Autistic Perception," *Trends in Cognitive Science* 16 (2012): 504–510. That paper was briefly responded to by Brock, J., "Alternative Bayesian Accounts of Autistic Perception: Comment on Pellicano and Burr," *Trends in Cognitive Sciences* 16(12) (2012): 573–574; and by Friston, K., Lawson, R., and Frith, C., "On Hyperpriors and Hypopriors: Comment on Pellicano and Burr," *Trends in Cognitive Sciences* 17(1) (January 2013): 504–505. The discussion continued in Van de Cruys, S., et al., "Precise Minds in Uncertain Worlds: Predictive Coding in Autism," *Psychological Review* 121(4) (October 2014): 649–675.

59 This favors an alternative theory: The work on Mooney images is by Cruys, S. Van de, et al., "The Use of Prior Knowledge for Perceptual Inference Is Preserved in ASD," *Clinical Psychological Science* 6(3) (2017): 382–393.

59 Converging evidence now favors: Further evidence for this was found using computational modeling and a trial-by-trial statistical learning paradigm. For more on this, and the experimental methods used to distinguish between the various options, see P. Karvelis, et al., "Autistic Traits, but Not Schizotypy, Predict Overweighting of Sensory Information in Bayesian Visual Integration," *eLife* 7 (2018): e34115. See also Palmer, C. J., Lawson, R. P., and Hohwy, J., "Bayesian Approaches to Autism: Towards Volatility, Action, and Behavior," *Psychological Bulletin* 143(5) (2017): 521–542.

59 "she feels in exquisite detail": The quoted passage is from George Musser's article "How Autism May Stem from Problems with Prediction," in *Spectrum*, March 7, 2018, https://www.spectrumnews.org/features/deep-dive/autism-may-stem-problems-prediction/.

60 Moreover, just as neurotypical people: The idea of autism spectrum condi-

tion as involving atypical niche construction is pursued in Constant, A., et al., "Precise Worlds for Certain Minds: An Ecological Perspective on the Relational Self in Autism," *Topoi* (2018): doi:10:1007/s11245-018-9546-4.

62 There are many videos online: A good online demonstration of the McGurk effect can be found on the BBC YouTube channel. For the other effect mentioned ("brainstorm" versus "green needle"), try https://www.youtube.com/watch?v=YvnOtS4V-Pg.

62 This makes sense if these individuals: See Zhang, J., et al., "McGurk Effect by Individuals with Autism Spectrum Disorder and Typically Developing Controls: A Systematic Review and Meta-Analysis," *Journal of Autism and Developmental Disorders* 49(1) (2019): 34–43.

62 In such cases, autism spectrum condition: A useful overview is provided by van Schalkwyk, G. I., Volkmar, F. R., and Corlett, P. R., "A Predictive Coding Account of Psychotic Symptoms in Autism Spectrum Disorder," *Journal of Autism and Developmental Disorders* 47 (2017): 1323–1340. See also Chouinard, P. A., et al., "The Shepard Illusion Is Reduced in Children with an Autism Spectrum Disorder Because of Perceptual Rather than Attentional Mechanisms," *Frontiers in Psychology* 9 (2018): 2452.

62 "I had to make sense, any sense": From Chadwick, P. K., "The Stepladder to the Impossible: A Firsthand Phenomenological Account of a Schizoaffective Psychotic Crisis," *Journal of Mental Health* 2 (1993): 239–250. The quote that follows is from page 239 of that article.

63 Important early research: See Fletcher, P., and Frith, C., "Perceiving Is Believing: A Bayesian Approach to Explaining the Positive Symptoms of Schizophrenia," *Nature Reviews Neuroscience* 10 (2009): 48–58. See also Corlett, P. R., Frith, C. D., and Fletcher, P. C., "From Drugs to Deprivation: A Bayesian Framework for Understanding Models of Psychosis," *Psychopharmacology* 206(4) (2009): 515–530; Corlett, P. R., et al., "Why Do Delusions Persist?," *Frontiers in Human Neuroscience* 3(12) (2009); and Corlett, P. R., et al., "Toward a Neurobiology of Delusions," *Progress in Neurobiology* 92(3) (2010): 345–369. For a useful review, see Griffin, J., and Fletcher, P., "Predictive Processing, Source Monitoring, and Psychosis," *Annual Review of Clinical Psychology* 13(1) (May 2017): 265–289.

64 Aberrant prediction errors, even if they play: There is, for example, emerging evidence that psychosis also involves diminished sensitivity to changes in the predictive value of sensory cues. This would further explain why psychotic symptoms, once present, are so remarkably resistant to counterevidence. See Powers, A. R., Mathys, C., and Corlett, P. R., "Pavlovian Conditioning-Induced Hallucinations Result from Overweighting of Perceptual Priors," *Science* 357(6351) (2017): 596–600. See also Corlett, P. R., Honey, G. D., and Fletcher, P. C., "Prediction Error, Ketamine and Psychosis: An Updated Model," *Journal of Psychopharmacology* (Oxford, UK), 30(11) (2016): 1145–1155; Teufel, C., et al., "Shift Toward Prior Knowledge Confers a Perceptual Advantage in Early Psychosis and Psychosis-Prone Healthy Individuals," *Proceedings of the National Academy of Sciences* 112(43) (2015): 13401–13406; and Sterzer, P., et al., "The Predictive Coding Account of Psychosis," *Biological Psychiatry* 84(9) (2018): 634–643.

64 But however complex the final story: A fuller picture of those checks

and balances will need to unravel the specific effects of impairments to systems involving different neurotransmitters such as dopamine, acetylcholine, serotonin, noradrenaline, and oxytocin. For a window onto that complex web of neuronal predictions and precision estimators, see Corlett, P., "Delusions and Prediction Error," in Bortolotti, L. (ed.), *Delusions in Context* (New York: Palgrave Macmillan, 2018), pp. 35–66.

65 Sometimes described as a "reality monitoring" deficit: A good general review of the neural circuitry of PTSD appears in Mahan, A. L., and Ressler, K. J., "Fear Conditioning, Synaptic Plasticity and the Amygdala: Implications for Posttraumatic Stress Disorder," *Trends in Neurosciences* 35(1) (2012): 24–35.

65 In some revealing recent experiments: The experiments are described in Homan, P., et al., "Neural Computations of Threat in the Aftermath of Combat Trauma," *Nature Reviews Neuroscience* 22 (March 2019): 470–476.

66 In the most severely affected individuals: The amygdala, an area known to be involved in the processing and prediction of fearful events, is both smaller in size and unusually active in those affected. Interestingly, in the affected veterans the amygdala and associated "fear" regions were actually less active in tracking the changing value of the cues. This leads the authors to speculate that the heightened impact of the prediction errors might itself be a kind of compensatory strategy emerging as the amygdala becomes less sensitive to the changing value of the cues. Cause and effect are here hard to unravel.

66 If this proves correct, tests like these: See Seriès, P., "Post-Traumatic Stress Disorder as a Disorder of Prediction," *Nature Reviews Neuroscience* 22 (March 2019): 329–336. An example of an experiment making use of other perceptual phenotyping techniques can be found in van Leeuwen, T. M., et al., "Perceptual Gains and Losses in Synesthesia and Schizophrenia," *Schizophrenia Bulletin* 47(3) (May 2021): 722–730.

67 To illustrate, one study: See Alderson-Day, B., et al., "Distinct Processing of Ambiguous Speech in People with Non-clinical Auditory Verbal Hallucinations," *Brain* 140(9) (2017): 2475–2489.

3. ACTION AS SELF-FULFILLING PREDICTION

71 predictive processing (active inference): As previously noted, I am using "predictive processing" to name the neurocomputational proposal also known in the literature as "active inference." Active inference is the term introduced by Karl Friston and colleagues as a way of highlighting the unity of perception and action under schemes in which perception aims to find the predictions that best fit the world while action aims to make the world (starting with simple bodily motions) fit the predictions. That deep unity is the topic of the present chapter. See Friston, K., "The Free-Energy Principle: A Unified Brain Theory?," *Nature Reviews Neuroscience* 11 (2010): 127–138; and Friston K. J., et al., "Action and Behavior: A Free-Energy Formulation," *Biological Cybernetics* 102 (2010): 227–260.

71 This became known as the "ideomotor theory of action": The classic works on the ideomotor theory are Lotze, H., *Medicinische Psychologie oder Physiologie der Seele* (Leipzig, Germany: Weidmannsche Buchhandlung,

1852); and James, W., *The Principles of Psychology*, Vols. I, II. (Cambridge: Harvard University Press, 1890/1950). For a handy contemporary review, Pezzulo, G., et al., "From Actions to Goals and Vice-Versa: Theoretical Analysis and Models of the Ideomotor Principle and TOTE," in Butz, M., et al. (eds.), *Anticipatory Behavior in Adaptive Learning Systems: Advances in Anticipatory Processing* (Springer, 2007), 73–93.

72 From the point of view of the marionette: For more on the marionette example, see Mohan, V., Bhat, A., and Morasso, P., "Muscleless Motor Synergies and Actions Without Movements: From Motor Neuroscience to Cognitive Robotics," *Physics of Life Reviews* 30 (2019): 89–111.

72 This is also known as the passive motion paradigm: See Mohan, V., and Morasso, P., "Passive Motion Paradigm: An Alternative to Optimal Control," *Frontiers in Neurorobotics* 5(4) (2011).

73 In the broadest possible terms, the solution: See Friston, K. J., and Parr, T., "Passive Motion and Active Inference," *Physics of Life Reviews* 30 (2019): 112–115.

74 The answer is again by means: For "generative models" see the Appendix.

75 In other words, I strongly predict: For a worked example, with all the computational steps spelled out, see Pio-Lopez, L., et al., "Active Inference and Robot Control: A Case Study," *Journal of the Royal Society Interface* (2016) 132016061620160616.

75 "proprioceptive" sensory information: See Cole, J., *Losing Touch: A Man Without His Body* (New York: Oxford University Press, 2016).

75 This is the predictive processing route to action: See Friston, K. J., et al., "Action and Behavior: A Free-Energy Formulation," *Biological Cybernetics* 102 (2010): 227–260.

76 Yet surprisingly enough, the wiring diagram of "motor cortex": For a review of the neurophysiology, see Shipp, S., "The Importance of Being Agranular: A Comparative Account of Visual and Motor Cortex," *Philosophical Transactions of the Royal Society B* 360 (2005): 797–814.

76 Predictive processing resolves this anomaly: See Shipp, S., Adams, R. A., and Friston K. J., "Reflections on Agranular Architecture: Predictive Coding in the Motor Cortex," *Trends in Neuroscience* 36 (2013): 706–716.

76 The surprising solution is for the brain: There is an excellent discussion of this in Brown, H., et al., "Active Inference, Sensory Attenuation and Illusions," *Cognitive Processing* 14(4) (2013): 411–427.

77 Even though no physical actions: For more on the relation between the ideomotor picture, sporting performance, and anticipatory imagination, see Koch, I., Keller, P., and Prinz, W., "The Ideomotor Approach to Action Control: Implications for Skilled Performance," *International Journal of Sport and Exercise Psychology* 2(4) (2004): 362–375.

78 That second copy, Holst believed: The 1950 paper was von Holst, E., and Mittelstaedt, H., "Das Reafferenzprinzip," *Naturwissenschaften* 37 (1950): 464–476. For a look at the complex history of these ideas, good places to start include Bruce Bridgeman's "A Review of the Role of Efference Copy in Sensory and Oculomotor Control Systems," *Annals of Biomedical Engineering* 23 (1995): 409–422; and Cullen, K., "Sensory Signals During Active Versus Passive Movement," *Current Opinion in Neurobiology* 14 (2004): 698–706.

78 Self-tickling is thus rather: For a wonderful exploration of the science and philosophy of humor and jokes, see Hurley, M. M., Dennett, D. C., and Adams Jr., R. B., *Inside Jokes: Using Humour to Reverse-Engineer the Mind* (Cambridge: MIT Press, 2011).

78 Under these bizarre conditions: An early exploration of self-tickling can be found in Weiskrantz, L., Elliot, J., and Darlington, C., "Preliminary Observations of Tickling Oneself," *Nature* 230(5296) (1971): 598–599. For a more detailed version, see Blakemore, S., Wolpert, D., and Frith, C., "Central Cancellation of Self-Produced Tickle Sensation," *Nature Neuroscience* 1(7) (1998): 635–640.

78 One reason why nervous systems: See, for example, Grush, R., "The Emulation Theory of Representation: Motor Control, Imagery, and Perception," *Behavioral and Brain Sciences* 27 (2004): 377–442.

79 The same situation arises in nuclear reactors: For a look at the links with applied areas such as bioreactor control, see Ungar, L., "A Bioreactor Benchmark for Adaptive Network-Based Process Control," in Miller, W., Sutton, R., and Werbos, P. (eds.), *Neural Networks for Control* (Cambridge: MIT Press, 1990).

79 Systems that instead predict: A recent framework that shares some of the flavor of Rick Grush's "emulator circuits" is due to Professor Larry Barsalou. See Barsalou, L. W., "Simulation, Situated Conceptualization, and Prediction," *Philosophical Transactions of the Royal Society B* 364(1521) (2009): 1281–1289.

79 Indeed, this is exactly what happens: See Kawato, M., Furukawa, K., and Suzuki, R. A., "Hierarchical Neural-Network Model for Control and Learning of Voluntary Movement," *Biological Cybernetics* 57 (1987): 169–185. See also Deuschl, G., et al., "Essential Tremor and Cerebellar Dysfunction: Clinical and Kinematic Analysis of Intention Tremor," *Brain* 123(8) (August 2000): 1568–1580.

80 But such dampening is indeed: Early results showing that simply expecting movement, regardless of how it is caused, leads to the dampening or attenuation of the sensory effects of those movements can be found in Voss, M., et al., "Mere Expectation to Move Causes Attenuation of Sensory Signals," *PLoS ONE* 3(8) (2008): e2866. For extensions and updates, see Kaiser, J., and Schütz-Bosbach, S., "Sensory Attenuation of Self-Produced Signals Does Not Rely on Self-Specific Motor Predictions," *European Journal of Neuroscience* 47(11) (2018): 1303–1310.

81 Vision itself, this body of work suggests: The active nature of vision (and of perception more generally) was at the core of much work in "ecological psychology"—for example, J. J. Gibson's classic work *The Ecological Approach to Visual Perception* (Boston: Houghton Mifflin, 1979). Work in "active vision" combined some of these insights with a rich computational and neurophysiological understanding, laying much of the groundwork for the fully integrated perception-action story outlined here. A classic early treatment in that field, which had a huge influence on my own thinking and work, was "A Critique of Pure Vision" by the neurophilosopher Patricia Churchland, the neuroscientist Vilayanur Ramachandran, and the computational neuroscientist Terence Sejnowski. This appeared in Koch, C.,

and Davis, J. (eds.), *Large-Scale Neuronal Theories of the Brain (*Cambridge: MIT Press, 1994), pp. 23–61.

81 This strategy provably affords: See Fink, P. W., et al., "Catching Fly Balls in Virtual Reality: A Critical Test of the Outfielder Problem," *Journal of Vision* 9(13) (2009): 14, 1–8. For a longer look at this example, see Chapter 8 of my 2016 book, *Surfing Uncertainty: Prediction, Action, and the Embodied Mind* (Oxford University Press).

81 It is also another example of controlling an action: This picture has deep affinities with an earlier model known as perceptual control theory (PCT). PCT argues that action-control systems control not what we do but what we sense. See Powers, W. T., *Behavior: The Control of Perception* (Chicago: Aldine de Gruyter, 1973). See also Mansell, W., and Carey, T. A., "A Perceptual Control Revolution?," *The Psychologist* 28 (November 2015): 896–899.

83 In the same way, the brain of the experienced tennis player: Fast-paced sports skills probably reflect a complex control strategy in which very direct and fast perception-action linkages combine with the ability to use changing context to make fluid, on-the-fly changes that respond intelligently to changing conditions. Understanding the interplay between these forms of knowledge and control is an important challenge for future work. For some helpful reflections on these general themes, see Sutton, J., "Batting, Habit and Memory: The Embodied Mind and the Nature of Skill," *Sport and Society* 10 (2007): 763–786.

86 At the bottom of all this lie predictions: See Friston, K. J., et al., "Deep Temporal Models and Active Inference," *Neuroscience and Biobehavioral Reviews* 77(6) (2017): 388–402.

87 We must at some level strongly predict: See Van de Cruys, S., Friston, K. J., and Clark, A., "Controlled Optimism: Reply to Sun and Firestone on the Dark Room Problem," *Trends in Cognitive Science* 24(9) (2020): 680–681.

87 We will then act in ways designed: There is lots more to say (and a long and technical literature that tries to say it) about just how this works and about how local decision making balances short-term and longer-term interests. See, for example, Friston, K., et al., "The Anatomy of Choice," *Philosophical Transactions of the Royal Society B* 369 (November 2014): 1655. For a fairly accessible treatment, see Pezzulo, G., Rigoli, F., and Friston, K., "Active Inference, Homeostatic Regulation and Adaptive Behavioural Control," *Progress in Neurobiology* 134 (November 2015): 17–35.

4. PREDICTING THE BODY

88 This is the so-called Dark Room puzzle: An early treatment of the Dark Room issues can be found in Friston, K., Thornton, C., and Clark, A., "Free-Energy Minimization and the Dark-Room Problem," *Frontiers in Psychology* 3 (2012): 1–7. A more recent treatment, covering a wider range of scenarios, is my 2018 piece, "A Nice Surprise? Predictive Processing and the Active Pursuit of Novelty," *Phenomenology and the Cognitive Sciences* 17(3) (2018): 521–534.

90 Even well-adapted darkness dwellers: The standard solution to the Dark Room puzzle—that we simply expect to be fed, to play, and to explore—appears briefly in Friston, K., "Embodied Inference: Or I Think Therefore

I Am, if I Am What I Think," in Tschacher, W., and Bergomi, C. (eds.), *The Implications of Embodiment (Cognition and Communication)* (Exeter, U.K.: Imprint Academic, 2011), pp. 89–125. But that solution is itself merely shorthand for a more complex and much more convincing story. See, for example, Schwartenbeck, P., et al., "Exploration, Novelty, Surprise, and Free Energy Minimization," *Frontiers in Psychology* 2013(4): 710.

90 This general idea has been traced to the nineteenth-century French physiologist: A useful review of the early history of the concept of homeostasis is Cooper, S. J., "From Claude Bernard to Walter Cannon. Emergence of the Concept of Homeostasis," *Appetite* 51(3) (2008): 419–427. See also Ramsay, D. S., and Woods, S. C., "Clarifying the Roles of Homeostasis and Allostasis in Physiological Regulation," *Psychological Review* 121(2) (2014): 225–247.

91 In the early days of cybernetics, self-regulating systems: Much of this was due to the influence of Norbert Wiener's 1948 MIT Press opus, *Cybernetics, or Control and Communication in the Animal and the Machine*. My tattered copy of that book proclaims itself, on the dust jacket, to be "A study of vital importance to psychologists, physiologists, electrical engineers, radio engineers, sociologists, philosophers, mathematicians, anthropologists, psychiatrists, and physicists." Though that list may at first sound wildly heterogenous, it would be entirely apt for today's emerging work on the predictive mind—though those electrical engineers and radio engineers might now be computer scientists and signal processing specialists.

91 There soon emerged a slightly more general concept: See Ramsay, D. S., and Woods, S. C., "Clarifying the Roles of Homeostasis and Allostasis in Physiological Regulation," *Psychological Review* 121(2) (2014): 225–247.

92 To make homeostasis and allostasis possible: Excellent introductions to interoception include A. D. Craig's "Interoception: The Sense of the Physiological Condition of the Body," *Current Opinion in Neurobiology* 13 (2003): 500–505; and Critchley, H. D., and Harrison, N. A., "Visceral Influences on Brain and Behavior," *Neuron* 77 (2003): 624–638. For an engaging longer treatment, there is Craig's 2016 book, *How Do You Feel? An Interoceptive Moment with Your Neurobiological Self* (Princeton University Press). Work by my Sussex colleagues Professor Hugo Critchley and Professor Sarah Garfinkel has been very influential in this area too, bringing these ideas into detailed contact with work on the predictive brain—for example, Critchley, H., and Garfinkel, S., "Interoception and Emotion," *Current Opinion in Psychology* 17 (2017): 7–14.

93 Estimations of error dynamics: The predictive processing account of this key tendency first appears in Joffily, M., and Coricelli, G., "Emotional Valence and the Free-Energy Principle," *PLoS Computational Biology* 9(6) (2013) e1003094.

93 Positive and negative moods: See Kiverstein, J., Miller, M., and Rietveld, E., "How Mood Tunes Prediction: A Neurophenomenological Account of Mood and Its Disturbance in Major Depression," *Neuroscience of Consciousness*, 2020, June 2; 2020(1):niaa003. See also (for the flip side of that) Miller, M., Rietveld, E., and Kiverstein, J., "The Predictive Dynamics of Happiness and Well-Being," *Emotion Review* 14(1) 2022: 15–30.

94 Within the sweet spot: The Goldilocks zone idea appears in Kidd, C., Piantadosi, S. T., and Aslin, R. N., "The Goldilocks Effect: Human Infants Allocate Attention to Visual Sequences That Are Neither Too Simple nor Too Complex," *PloS ONE* 7(5) (January 2012): e36399.

94 This was especially true in simulated environments: For the work in artificial curiosity, see Oudeyer, P., and Smith, L. B., "How Evolution May Work Through Curiosity-Driven Developmental Process," *Topics in Cognitive Science* 8 (2016): 492–502. There's a good introduction by John Pavlus online in the September 2017 issue of *Quanta* magazine too: https://www.quantamagazine.org/clever-machines-learn-how-to-be-curious-20170919/.

95 Instead, they will constantly seek out: This tendency has a dark side too. It has recently been speculated that such creatures, in virtue of the way biological systems implement these tendencies, become especially vulnerable to certain drugs of addiction. These drugs hijack that same underlying mechanism, making our brains estimate—falsely—that we are suddenly doing better than expected, and thus rendering drug use a deeply attractive action, drawing us back again and again. See for example Miller, M., Kiverstein, K., and Rietveld, E., "Embodying Addiction: A Predictive Processing Account," *Brain and Cognition* 138 (2020).

96 "Every thought, memory, emotion": The quoted passage is from Barrett, L. F., *How Emotions Are Made: The Secret Life of the Brain* (UK: Pan Macmillan, 2017), p. 121.

97 However, no such simple analogs exist: It's probably impossible to prove this negative, but the evidence against it is solid—see Critchley, H. D., "Neural Mechanisms of Autonomic, Affective, and Cognitive Integration," *The Journal of Comparative Neurology* 493(1) (2005): 154–66.

97 On the contrary, large and convincing studies: See for example Siegel, E. H., et al., "Emotion Fingerprints or Emotion Populations? A Meta-Analytic Investigation of Autonomic Features of Emotion Categories," *Psychological Bulletin* 144(4) (2018): 343–393.

97 Instead, emotional experience seems to be constructed: For a lively and comprehensive account of emotion as constructed from these melting pots of influence, see Barrett, *How Emotions Are Made*.

98 This takes us way beyond: Interoceptive predictive processing has been explored in rich physiological and neurophysiological detail in a series of publications by Professor Lisa Feldman Barrett and colleagues on what they dub the EPIC (Embodied Predictive Interoception Coding) model—see Barrett, L. F., and Simmons, K., "Interoceptive Predictions in the Brain," *Nature Reviews Neuroscience* 16(7) (July 2015): 1–11. For a largely complementary picture but with a more "cybernetic" twist, see work by my University of Sussex colleague Professor Anil Seth—for example, "Interoceptive Inference, Emotion, and the Embodied Self," *Trends in Cognitive Sciences* 17(11) (November 2013). For something more formal, try Pezzulo, G., Rigoli, F., and Friston, K., "Active Inference, Homeostatic Regulation and Adaptive Behavioural Control," *Progress in Neurobiology* 134 (November 2015): 17–35.

99 It has been centrally implicated: For example, Craig, A. D., "How Do You

Feel—Now? The Anterior Insula and Human Awareness," *Nature Reviews Neuroscience* 10(1) (2009): 59–70.

99 We can contrast this picture: See, e.g., Scherer, K. R., "The Dynamic Architecture of Emotion: Evidence for the Component Process Model," *Cognition and Emotion* 23(7) (2009): 1307–1351. For a general review, see Scherer, K. R., "Appraisal Theory," in Dalgleish, T., and Power, M. J. (eds.), *Handbook of Cognition and Emotion* (New York: Guilford Press, 1999), pp. 637–663.

101 In just this way, the great Russian physiologist: See, for example, Pavlov, I. P., *Lectures on Conditioned Reflexes: Twenty-five Years of Objective Study of the Higher Nervous Activity (Behaviour) of Animals*, W. H. Gantt, trans. (New York: International Publishers, 1928).

101 It has in fact long been part: For a compelling argument that the linear model needs to be abandoned, see Luis Pessoa's *The Cognitive Emotional Brain: From Interactions to Integration* (Cambridge: MIT Press, 2013). Contemporary neuroscientist Joseph LeDoux charts the history of ideas about the physiological basis of emotion in his entry on "Emotion," in Plum, F. (ed.), *Handbook of Physiology 1: The Nervous System*. Vol. 5, *Higher Functions of the Brain* (Bethesda, M.D.: American Physiological Society, 1987), pp. 419–460. My own thinking on these topics has also been greatly influenced by conversations and collaborations with Dr. Mark Miller.

101 What resulted is a complex looping arrangement: This is sometimes expressed by talk of "processes of continuous reciprocal causation" that bind multiple components into unified dynamic wholes. Much of my own earlier work involved looking long and hard at the conceptual consequences of these kinds of looping patterns of influence, both within the brain and in larger brain-body-world systems. The fullest expression of this picture is in *Being There: Putting Brain, Body, and World Together Again* (Cambridge: MIT Press, 1997). Patterns of continuous reciprocal causal influence are part of the subject matter of a large body of work, with applications all across the sciences, known as "dynamical systems theory."

101 It is this looping circuitry: On this dense causal web, see Sherman. S. M., and Guillery, R. W., "Distinct Functions for Direct and Transthalamic Corticocortical Connections," *Journal of Neurophysiology* 106 (2011): 1068–1077, 1073. For a powerful exploration of the pivotal role of the thalamus, see Sherman, S. M., and Guillery, R. W., *Thalamocortical Processing: Understanding the Messages That Link the Cortex to the World* (Cambridge: MIT Press, 2013). Ray Guillery died in April 2017. His solo-authored book, *The Brain as a Tool* (Oxford University Press), was published posthumously in October that same year, and offers a wonderfully accessible treatment of the core idea that sensory and motor information are constantly combined during neural processing and bodily action.

102 The moment-by-moment control of action: See Parvizi, J., "Corticocentric Myopia: Old Bias in New Cognitive Sciences," *Trends in Cognitive Sciences* 13(8) (2009): 354–359.

102 These cortico-subcortical loops: See Kanai, R., et al., "Cerebral Hierarchies: Predictive Processing, Precision and the Pulvinar," *Philosophical Transactions of the Royal Society B*, 370: 20140169.

102 Because so many subcortical circuits: For this "feed-around" picture,

accompanied by a fully detailed physiological plot, see Lewis, M., "Bridging Emotion Theory and Neurobiology Through Dynamic Systems Modelling," *Behavioral and Brain Sciences* 28 (2005): 169–245.

102 Intriguingly, whenever the experimenters induced: See Valins, S., "Cognitive Effects of False Heart-Rate Feedback," *Journal of Personality and Social Psychology* 4 (1996): 400–408.

103 The experimenters found that when the false feedback: See Gray, M. A., et al., "Modulation of Emotional Appraisal by False Physiological Feedback During fMRI," *PLoS ONE* 2(6) (2007): e546.

103 The predictive brain then treats: See Anderson E., et al., "Out of Sight but Not Out of Mind: Unseen Affective Faces Influence Evaluations and Social Impressions," *Emotion* 12 (2012): 1210–1221.

104 Perhaps the person's smile now seems: The speculations on Capgras delusion can be found in Griffin, J. D., and Fletcher, P. C., "Predictive Processing, Source Monitoring, and Psychosis," *Annual Review of Clinical Psychology* 13(1) (2017): 265–289.

104 This plausibly sets the scene: In such scenarios, beliefs and perceptual experiences again get locked into a mutually supportive but profoundly misleading cycle—one that has usefully been dubbed "circular belief propagation." See Jardri, R., and Denève, S., "Circular Inferences in Schizophrenia," *Brain* 136(11) (2014): 3227–3241.

105 This unifying perspective: See for example Swardfager, W., et al., "Mapping Inflammation onto Mood: Inflammatory Mediators of Anhedonia," *Neuroscience and Biobehavorial Review* 64 (2016): 148–166; Gold, P. W., "The Organization of the Stress System and Its Dysregulation in Depressive Illness," *Molecular Psychiatry* 20 (2015): 32–47.

105 Tying all these together: See Barrett, L. F., Quigley, K. S., and Hamilton, P., "An Active Inference Theory of Allostasis and Interoception in Depression," *Philosophical Transactions of the Royal Society B* 371 (2016): 20160011.

108 They may also take more general forms: See Rief, W., and Joormann, J., (2019). "Revisiting the Cognitive Model of Depression: The Role of Expectations," *Clinical Psychology in Europe* 1(1): 1–19.

108 Where healthy controls rapidly: Kube, T., et al., "Why Dysfunctional Expectations in Depression Persist—Results from Two Experimental Studies Investigating Cognitive Immunization," *Psychological Medicine* 49(9) (2019): 1532–1544. See also Kube T., et al., "Distorted Cognitive Processes in Major Depression: A Predictive Processing Perspective," *Biological Psychiatry* 87 (2020): 388–398. For some reservations and questions, see Harmer, C. J., and Browning, M., "Can a Predictive Processing Framework Improve the Specification of Negative Bias in Depression?," *Biological Psychiatry* 87(5) (March 1, 2020): 382–383.

109 In all these cases, predictive processing: Major players here include the neurotransmitters dopamine and acetylcholine—see, for example, Schwartenbeck, P., et al., "The Dopaminergic Midbrain Encodes the Expected Certainty About Desired Outcomes," *Cerebral Cortex* 25 (2015): 3434–3445; and Diaconescu, A. O., et al., "Hierarchical Prediction Errors in Midbrain and Septum During Social Learning," *Social Cognitive and Affective Neuroscience* 12 (2017): 618–634.

109 This is also what makes the new theories testable: For this suggestion, and a useful review of this general area, see Smith, R., Badcock, P., and Friston, K. J., "Recent Advances in the Application of Predictive Coding and Active Inference Models Within Clinical Neuroscience," *Psychiatry and Clinical Neurosciences* 75 (2021): 3–13.

110 Aesthetic chills occur in many contexts: There is a large and varied literature on this. A good place to start is with Schoeller, F., "The Shivers of Knowledge," *Human Social Studies* 4(3) (2015): 44–55. See also Goldstein, A., "Thrills in Response to Music and Other Stimuli," *Physiological Psychology* 8(1) (1980): 126–129.

110 This is because music is a domain: There is a fascinating literature emerging on music and the predictive brain. See Koelsch, S., Vuust, P., and Friston, K., "Predictive Processes and the Peculiar Case of Music," *Trends in Cognitive Sciences* 23(1) (January 2019): 63–77.

111 Recent work explores this idea: See Lehne, M., and Koelsch, S., "Toward a General Psychological Model of Tension and Suspense," *Frontiers in Psychology* 6 (2015): 79.

111 "probability designs": Kukkonen applies her account to literary narratives such as novels, poems, and short stories. Literary narratives set up, manipulate, and resolve expectations rather like the design of a roller coaster, or a piece of music. They build anticipations and deliver resolutions (that can set up new anticipations, and generate new uncertainties, in turn). See Kukkonen, K., *Probability Designs: Literature and Predictive Processing* (UK: Oxford University Press, 2019).

112 "Frisson prosthesis": See Felix Schoeller, F., et al., "Enhancing Human Emotions with Interoceptive Technologies," *Physics of Life Reviews* 31 (2019): 310–319.

113 The Frisson prosthesis acts in the same basic way: For some updates on the predictive processing account of aesthetic chills, see Miller, M., et al., "Getting a Kick Out of Film: Aesthetic Pleasure and Play in Prediction Error Minimizing Agents," in *Worldling the Brain*, forthcoming.

113 This enables living beings to bring forth: This talk of "bringing forth" a world of mattering and importance is referencing what have become known as "enactive" approaches to the study of mind and adaptive behavior—see, e.g., Varela, F., Thompson, E., & Rosch, E., *The Embodied Mind* (Cambridge: MIT Press, 1991). There are hints of such a picture in Merleau-Ponty, M., *The Phenomenology of Perception* (Colin Smith, trans.) (London: Routledge & Kegan Paul, 1945/1962). For some discussion of the similarities and differences between the predictive processing account and these pioneering works, see Chapter 9 of my 2016 Oxford University Press book, *Surfing Uncertainty: Prediction, Action, and the Embodied Mind*. See also Gallagher, S., and Allen, M., "Active Inference, Enactivism and the Hermeneutics of Social Cognition," *Synthese* 195 (2018): 2627–2648.

INTERLUDE: THE HARD PROBLEM—PREDICTING THE PREDICTORS?

115 What, you may ask, does all this tell us: Classic papers about this puzzle include Nagel, T., "What Is It Like to Be a Bat?," *Philosophical Review* 83(4)

(October 1974): 435–456; and Levine, J., "Materialism and Qualia: The Explanatory Gap," *Pacific Philosophical Quarterly* 64(4) (October 1983): 354–361.

116 "hard problem of consciousness": For a long and careful examination of the "hard problem" (though one that comes to some very different conclusions to our own), see Chalmers, D., *The Conscious Mind: In Search of a Fundamental Theory* (New York: Oxford University Press, 1996).

116 "meta-problem of consciousness": For more on this, see Chalmers, D., "The Meta-Problem of Consciousness," *Journal of Consciousness Studies* 25(9–10)(2018): 6–61. There is a subsequent special issue dedicated to the meta-problem topic—*Journal of Consciousness Studies* 26(9–10) (2019).

117 This is because they aim: Such an approach is most famously deployed by Daniel Dennett, in *Consciousness Explained* (Boston: Little, Brown, 1991). In recent years it has often taken the form of a defense of "illusionism." I think the name is unfortunate, but the key ideas are correct. The best defense of illusionism is probably Frankish, K., "Illusionism as a Theory of Consciousness," *Journal of Consciousness Studies* 23(11–12) (2016): 11–39, where it appears along with multiple replies and commentaries.

117 For example, we just saw that interoceptive sensory information: Such interactions have been powerfully demonstrated in experiments that present neutral and fearful stimuli while manipulating self-estimated heart rate, showing that neutral stimuli are more often seen as fearful when heart rate is estimated as increased. For these experiments and many more, see Chapter 4.

118 Philosophers and psychologists talk here of "affordances": A useful paper on affordances, that draws direct links with predictive processing accounts, is Bruineberg, J., and Rietveld, E., "Self-Organization, Free Energy Minimization, and Optimal Grip on a Field of Affordances," *Frontiers in Human Neuroscience* 8 (2014): 599.

118 Emotion—or so we argued: For more on error dynamics, see Kiverstein, J., Miller, M., and Rietveld, E., "The Feeling of Grip: Novelty, Error Dynamics, and the Predictive Brain," *Synthese* 196 (2019): 2847–2869.

118 It is a kind of marker: See Miller, M., Kiverstein, J., and Rietveld, E., "The Predictive Dynamics of Happiness and Well-Being," *Emotion Review* 14(1) (2022): 15–30. See also Kiverstein, J., Miller, M., and Rietveld, E., "How Mood Tunes Prediction: A Neurophenomenological Account of Mood and Its Disturbance in Major Depression," *Neuroscience of Consciousness* 2020(1), Article niaa003.

118 In response, the brain increases its learning rate: For more on the role of learning rates in predictive inference, see Hohwy, J., "Priors in Perception: Top-Down Modulation, Bayesian Perceptual Learning Rate, and Prediction Error Minimization," *Consciousness and Cognition* 47 (2017): 75–78.

119 The term sentience: Burns, J. H., and Hart, H. L. A. (eds.), *The Collected Works of Jeremy Bentham: An Introduction to the Principles of Morals and Legislation* (Oxford: Oxford University Press, 1970).

120 Expecting Ourselves: Many of the ideas presented here—and throughout this book—took shape while I was Principal Investigator on a large European Research Council Advanced Grant called Expecting Ourselves:

Embodied Prediction and the Construction of Conscious Experience (XSPECT—DLV-692739). You can find out more about that work at https://www.x-spect.org/.

120 A recurrent theme, one that kept nagging: This idea was mooted by Dennett, first at the farmhouse and then in a short commentary on my paper "Whatever Next? Predictive Brains, Situated Agents, and the Future of Cognitive Science." It appeared, along with multiple helpful and challenging commentaries, in *Behavioral and Brain Sciences* 36(3) (2013): 181–204. Dennett's commentary was called "Expecting Ourselves to Expect: The Bayesian Brain as a Projector," and appears on pages 209–210 of the same issue.

120 the "strange inversion": See, for example, Dennett, D., "Why and How Does Consciousness Seem the Way It Seems?," in Metzinger, T., and Windt, J. M. (eds.), *Open MIND* (Frankfurt am Main: MIND Group, 2015).

123 Once in command of a structured predictive model: For more on that important notion, see the Appendix.

123 When new inputs are swept under: See Clark, A., "Strange Inversions: Prediction and the Explanation of Conscious Experience," in Huebner, B. (ed.), *Engaging Daniel Dennett* (Oxford University Press, 2016).

125 This is how we "predict the predictors": For a very compatible account, couched in terms of simplified models of our own attentional processing, see Graziano, M. *Rethinking Consciousness: A Scientific Theory of Subjective Experience* (New York: W. W. Norton, 2019). See also Hoffman, D. D., Singh, M., and Prakash, C., "The Interface Theory of Perception," *Psychonomic Bulletin & Review* 22 (2015): 1480–1506.

126 His reply was that such an agent: See Chalmers, D., *The Conscious Mind* (Oxford University Press, 1996), p. 291.

126 Chalmers himself has frequently invoked: For a balanced discussion of the philosophical zombie thought experiment, see Kirk, R., "Zombies," in *The Stanford Encyclopedia of Philosophy* (Spring 2019 edition), https://plato.stanford.edu/archives/spr2019/entries/zombies/.

127 Instead, we should be using a notion: For more on this important constraint, see Chalmers, D., "The Meta-Problem of Consciousness," *Journal of Consciousness Studies* 25(9–10) (2018): 6–61.

128 Qualitative consciousness is real: For a fuller version of these admittedly impressionistic arguments, see Clark, A., "Consciousness as Generative Entanglement," *Journal of Philosophy* 116(12) (2019): 645–662; and Clark, A., Friston, K., and Wilkinson, S., "Bayesing Qualia: Consciousness as Inference, Not Raw Datum," *Journal of Consciousness Studies* 26 (9–10) (2019): 19–33.

5. EXPECTING BETTER

129 "all manner of great and small vessels": The quoted passage is from a letter reproduced in Friedman, D. M., *A Mind of Its Own: A Cultural History of the Penis* (New York: Free Press, 2002), pp. 76–77. For the full story, try Pinto-Correia, C., *The Ovary of Eve* (Chicago: University of Chicago Press, 1997).

129 The story about Leeuwenhoek: I first encountered the example of Leeuwenhoek in Susanna Siegel's wonderful book *The Rationality of Per-*

ception (New York: Oxford University Press, 2017)—though Siegel suggests it may well be historically inaccurate, or even apocryphal.

130 A simple example, lightly adapted: The example of Angry Jack is from the same book by Susanna Siegel—*The Rationality of Perception*.

131 In some of those experiments, false cardiac feedback: The work involving false cardiac feedback is Gray, M. A., et al., "Modulation of Emotional Appraisal by False Physiological Feedback During fMRI," *PLoS ONE* 2 (6) (2207): e546.

131 A recent paper looking at shooter bias: The figures are drawn from Fachner, G., and Carter, S., "An Assessment of Deadly Force in the Philadelphia Police Department" (Washington, D.C.: Collaborative Reform Initiative, Office of Community Oriented Policing Services, 2018), pp. 1–173.

132 The bodily sensations themselves, Barrett notes: The police officer scenario is adapted from Fridman, J., et al., "Applying the Theory of Constructed Emotion to Police Decision Making," *Frontiers in Psychology* 10 (2019): 1946.

132 Similarly, an individual who has just experienced anger: See Baumann, J., and DeSteno, D., "Emotion Guided Threat Detection: Expecting Guns Where There Are None," *Journal of Personality and Social Psychology* 99 (4) (2010): 595–610.

132 The upshot is that, to use an evocative phrase: See Anderson, E., Siegel, E. H., and Barrett, L. F., "What You Feel Influences What You See: The Role of Affective Feelings in Resolving Binocular Rivalry," *Journal of Experimental Social Psychology* 47 (2011): 856–860.

132 This is just another manifestation of that continuous line: There are also micro-versions of such effects that emerge moment by moment, as our own heartbeat traces its ongoing path through systole, on the heartbeat, the moment when the heart pushes out blood, and diastole, the island of calm between those more explosive moments. Unsurprisingly, physiological arousal is greatest at systole (on the heartbeat) and this information flows to the brain via baroreceptors, which are pressure sensors responding to changes in tension at the arterial wall. It turns out that fear-inducing stimuli presented at systole are enhanced, and have greater impacts than those same stimuli if presented between the heartbeats, at diastole. Recent experimental work shows that Black people are more often mis-seen as holding weapons than White people when the target images are presented on the systole rather than on the diastole. See Azevedo, R. T., et al., "Cardiac Afferent Activity Modulates the Expression of Racial Stereotypes," *Nature Communications* 8 (2017): 13854.

133 The influence of media depictions: The studies appear in Wormwood, J. B., et al., "Threat Perception After the Boston Marathon Bombings: The Effects of Personal Relevance and Conceptual Framing," *Cognition and Emotion* 30 (2016): 539–549.

133 It is well known that the many ways: In another study, people who had previously shared fake political news online were asked to rate the accuracy of some nonpolitical headlines. Having engaged in this simple exercise, they became less likely to share fake political news in a period following the exercise. This suggested to the researchers that much sharing of

fake news online is not malicious so much as simply premature—a kind of social media reflex reaction that can be partially offset by engaging in a few simple practices which remind us that news headlines can be fake or inaccurate. This breaks the self-confirming cycle by dampening the unreflective shares that feed and maintain it. See Pennycook, G., et al., "Understanding and Reducing the Spread of Misinformation Online" (November 13, 2019), https://doi.org/10:31234/osf.io/3n9u8.

134 In cases such as these, it is the cultural milieu: For discussion, see Hinton, P., "Implicit Stereotypes and the Predictive Brain: Cognition and Culture in 'Biased' Person Perception," *Palgrave Communications* 3 (2017): 86. The figures for male recruits to engineering are from the American Society of Mechanical Engineers (ASME) website, in a piece by Mark Crawford entitled "Engineering Still Needs More Women"—https://www.asme.org/topics-resources/content/engineering-still-needs-more-women.

135 They may depict misinformed or unhelpful: More subtly, yet equally perniciously, they may simply fail to depict core characters with disabilities, or to feature any non-White leading characters. Indeed, according to a study by the CLPE (a U.K. body—the Centre for Literacy in Primary Education) only 1 percent of children's books published in the U.K. in 2017 featured a BAME main character—BAME being a U.K. demographic meaning Black, Asian, and Minority Ethnic. And only 4 percent included BAME background characters. The CLPE survey figures are reported by Bold, M. R., et al., "Representation of People of Color Among Children's Book Authors and Illustrators"—available at www.booktrust.org.uk/globalassets/resources/represents/booktrust-represents-diversity-childrens-authors-illustrators-report.pdf.

135 We can act to remedy this: Examples might include the emerging subgenre of positive science fiction, as exemplified by Becky Chambers's *Wayfarers* (aka Murderbot) trilogy, now published with Hodder & Stoughton. These books combine engaging plots with compelling depictions of complex multispecies cooperation. Equally salutary are the profound racial and historical alterations and reversals depicted in the BBC series *Noughts & Crosses*. Of course, we also need works such as Margaret Atwood's *The Handmaid's Tale*, which help focus attention on horrors, atrocities, and injustices. There are no simple answers here, but better understanding the role and potential importance of fictional worlds as training arenas for the predictive brain is a good first step.

135 Later, when asked to estimate the size: See Keizer, A., et al., "A Virtual Reality Full Body Illusion Improves Body Image Disturbance in Anorexia Nervosa," *PLoS ONE* 11(10) (2016): e016392. See also Gadsby, S., "Manipulating Body Representations with Virtual Reality: Clinical Implications for Anorexia Nervosa," *Philosophical Psychology* 32:6 (2019): 898–922. For an application of predictive processing to understanding anorexia, see Gadsby, S., and Hohwy, J., "Why Use Predictive Processing to Explain Psychopathology? The Case of Anorexia Nervosa," in S. Gouveia, R. Mendonça, and M. Curado (eds.), *The Philosophy and Science of Predictive Processing* (London: Bloomsbury, 2020).

136 Getting action safely into the training circuit: See, for example, Lamb, H.,

"Good Cop, Good Cop: Can VR Help to Make Policing Kinder?," *Engineering and Technology* (January 8, 2020).

136 This is already being achieved: Examples of such trial programs are discussed in Arnetz, B. B., et al., "Assessment of a Prevention Program for Work-Related Stress Among Urban Police Officers," *International Archives of Occupational and Environmental Health* 86 (2013): 79–88; Arnetz, B. B., et al., "Trauma Resilience Training for Police: Psychophysiological and Performance Effects," *Journal of Police and Criminal Psychology* 24 (2009): 1–9; and Andersen, J. P., and Gustafsberg, H., "A Training Method to Improve Police Use of Force Decision Making: A Randomized Controlled Trial," *SAGE Open* 6 (2016): 1–13.

136 But systematic training regimes: See Andersen, J. P., et al., "Reducing Lethal Force Errors by Modulating Police Physiology," *Journal of Occupational and Environmental Medicine* 60 (2018): 867–874.

136 Such training, as Barrett notes: For the data on police officer health, see Violanti, J. M., et al., "Posttraumatic Stress Symptoms and Subclinical Cardiovascular Disease in Police Officers," *International Journal of Stress Management* 13 (2006): 541–544.

137 This means that you are more likely: See Quadt, L., Critchley, H. D., and Garfinkel, S. N., "The Neurobiology of Interoception in Health and Disease," *Annals of the New York Academy of Sciences* 1428(1) (2018): 112–128. There is, however, considerable debate concerning the validity of many current measures of interoceptive accuracy. See Ring, C., and Brener, J., "Heartbeat Counting Is Unrelated to Heartbeat Detection: A Comparison of Methods to Quantify Interoception," *Psychophysiology* 55(9) (September 2018): e13084.

137 Inaccurate and "coarse" information: The picture is quite complex though, as anxiety may result from an overemphasis on sensory evidence against expectations. Emerging evidence suggests that this is in turn linked to disturbances involving the interoceptive and body budgeting networks discussed in the previous chapter. See Barrett, L. F., *How Emotions Are Made* (Pan Macmillan, 2017), Chapter 10.

137 Garfinkel speculates that this extreme self-accuracy: The comments about the hostage negotiator appear in the online magazine *The Psychologist* 32 (January 2019): 38–41, under the banner "It's an intriguing world that is opening up"—see https://thepsychologist.bps.org.uk/volume-32/january-2019/its-intriguing-world-opening.

138 In this work, participants with better interoceptive self-awareness: See Mulcahy, J. S., et al., "Interoceptive Awareness Mitigates Deficits in Emotional Prosody Recognition in Autism," *Biological Psychology* 146 (2019): 107711.

138 For example, Fig. 5.2 is the famous logo from the 2014 FIFA World Cup: I borrow this example from Madrigal, A., "Things You Cannot Unsee (and What They Say About Your Brain)," *The Atlantic* (May 5, 2014).

142 To return to a metaphor used in Chapter 1: What about the world as revealed by science? As you might expect, this is something of a metaphysical minefield—one that lies far beyond the scope of this short treatment. A good place to start is *Theory and Reality: An Introduction to the Philosophy*

of Science, by Peter Godfrey-Smith (Chicago: University of Chicago Press, second edition, 2021).

142 The experienced world was like the message: See Merleau-Ponty, M., *The Phenomenology of Perception* (Colin Smith, trans.) (London: Routledge & Kegan Paul, 1945/1962). See also Varela, F., Thompson, E., and Rosch, E., *The Embodied Mind* (Cambridge: MIT Press, 1991). I further discuss this example and the issues it raises in Chapter 9 of *Surfing Uncertainty* (Oxford University Press, 2016).

143 Is that closer to, or further from: But in the kinds of social and cultural worlds neurotypical humans have constructed, the altered balances found in ASC can act as a serious barrier to learning and social fluency. When the "volume" on the incoming sensory signal is turned up even small prediction errors count as salient. That persistent unresolved error breeds anxiety and leads to various workarounds (such as controlling the environment by whatever means available). For more on anxiety, depression, and the predictive brain, see Chapter 4. See also Smith, R., Badcock, P., and Friston, K. J., "Recent Advances in the Application of Predictive Coding and Active Inference Models Within Clinical Neuroscience," *Psychiatry and Clinical Neurosciences* 75 (2021): 3–13.

143 My University of Sussex colleague Professor Anil Seth: See Seth, A., "The Neuroscience of Reality," *Scientific American* 321(3) (2019): 40–47. The piece ran with the evocative tagline "Reality is constructed by the brain, and no two brains are exactly alike."

144 Some of what works best: See the brief discussion of the important notion of "amortized inference" in Chapter 1, ninth note, and again in the Appendix, seventh note.

144 So human experience reflects: See Teufel, C., and Fletcher, P., "Forms of Prediction in the Nervous System," *Nature Reviews Neuroscience* 21 (2020): 231–242. But even processing that relies heavily on fixed structural constraints may often be highlighted or suppressed by varying estimates of precision, since these determine which neural responses are to be given the greatest weight as we perform some task. To get a sense of the way this kind of distinction (between structural constraints and flexible predictions) falls into place within the even larger framework known as "free energy minimization," see Friston, K., and Buzsáki, G., "The Functional Anatomy of Time: What and When in the Brain," *Trends in Cognitive Sciences* 20(7) (July 2016): 500–511.

145 Such effects, especially in time-pressured: I am not suggesting that all cases of shooter bias are rooted in misperceptions of this kind. The point, rather, is that these kinds of effect are real and may at times have contributed to such events.

6. BEYOND THE NAKED BRAIN

147 "I rely on apps such as SwiftKey": The quote is from Goldstaub, T., "How Artificial Intelligence Helped Me Overcome My Dyslexia," *The Guardian* (December 13, 2020).

150 One paper in the volumes: The paper was Rumelhart, D. E., et al., "Schemata and Sequential Thought Processes in Parallel Distributed Process-

ing," in Rumelhart, D. E., McClelland, J. L., and the PDP Research Group, *Parallel Distributed Processing: Explorations in the Microstructure of Cognition*, Vol. 2, *Psychological and Biological Models* (Cambridge: MIT Press, 1986), pp. 7–57.

151 Understanding this process reveals: See, for example, my 1997 MIT Press book, *Being There: Putting Brain, Body, and World Together Again*. In many ways this is my favorite book-child, joyfully riding multiple waves of excitement about the embodied mind.

152 Nowadays, there is widespread recognition: See, for example, https://www.enablingenvironments.com.au/.

153 Seen from that angle we are all indeed cyborgs: See especially my 2008 book, *Supersizing the Mind: Action, Embodiment and Cognitive Extension* (New York: Oxford University Press); and my 2003 popular treatment, *Natural-Born Cyborgs: Minds, Technologies, and the Future of Human Intelligence* (New York: Oxford University Press).

153 The damage can be somewhat repaired: For an excellent review of the state of the art, see Calkins, M. P., "From Research to Application: Supportive and Therapeutic Environments for People Living with Dementia," *Gerontologist* 58 (Suppl 1) (January 2018): S114–S128; and Holthe, T., et al., "Usability and Acceptability of Technology for Community-Dwelling Older Adults with Mild Cognitive Impairment and Dementia: A Systematic Literature Review," *Clinical Interventions in Aging* 13 (May 4, 2018): 863–866. For a look at these issues from the perspective of work on the extended mind, see Drayson, Z., and Clark, A., "Cognitive Disability and Embodied, Extended Minds," in Cureton, A., and Wasserman, D. T. (eds.), *The Oxford Handbook of Philosophy and Disability* (New York: Oxford University Press, 2020).

155 The experts select and lay out: The bartender experiments appear in Beach, K., "The Role of External Mnemonic Symbols in Acquiring an Occupation," in Gruneberg, M. M., and Sykes, R. N. (eds.), *Practical Aspects of Memory* (New York: Wiley, 1988), pp. 342–346.

155 In classic research from the late 1990s: See Kirsh, D., and Maglio, P., "On Distinguishing Epistemic from Pragmatic Action," *Cognitive Science* 18 (1994): 513–549; and Kirsh, D., and Maglio, P., "Reaction and Reflection in Tetris," in Hendler, J. (ed.), *Artificial Intelligence Planning Systems: Proceedings of the First Annual Conference AIPS* (San Mateo, CA: Morgan Kaufman, 1992).

156 This is sometimes called the "coastal navigation algorithm": Roy, N., and Thrun, S., "Coastal Navigation with Mobile Robots," in *Advances in Neural Information Processing Systems 12* (Cambridge: MIT Press, 2000). See also Pezzulo, G., and Nolfi, S., "Making the Environment an Informative Place: A Conceptual Analysis of Epistemic Policies and Sensorimotor Coordination," *Entropy* 21(4) (2019): 350; https://doi.org/10:3390/e21040350.

156 Orangutans are famously adept tool users: See Jabr, F., "An Orangutan Learns to Fish," *The New Yorker* (Annals of Technology) (September 17, 2014).

157 Fig. 6.1 Mego the orangutan: The pictures are from https://www.dailymail.co.uk/news/article-2746844/Don-t-depth-Incredible-pictures

-orangutan-using-stick-check-river-safe-cross.html. It is possible, of course, that Mego was not actually testing for depth. I use the case merely as a colorful illustration. But there is no doubt that many nonhuman animals perform epistemic actions of various kinds. For a scholarly treatment of the capacities of orangutans, see Laumer, I. N., et al., "Orangutans (*Pongo abelii*) Make Flexible Decisions Relative to Reward Quality and Tool Functionality in a Multi-Dimensional Tool-Use Task," *PLoS ONE* 14(2) (2019): e0211031; DOI: 10:1371/journal.pone.0211031.

158 Actions are then chosen that deliver: The deep unity between practical and epistemic action is further explored in Donnarumma, F., et al., "Action Perception as Hypothesis Testing," *Cortex* 89 (2017): 45–60; and in Pezzulo, G., and Nolfi, S., "Making the Environment an Informative Place: A Conceptual Analysis of Epistemic Policies and Sensorimotor Coordination," *Entropy* 21 (2019): 350. For some interesting work on the neural signatures of actions that improve information in monkeys, see Foley, N. C., et al., "Parietal Neurons Encode Expected Gains in Instrumental Information," *Proceedings of the National Academy of Sciences of the United States of America* 2017: 114 (16) E3315-E3323.

159 The simulated rats started each run: The rats (small bundles of code) were set up to expect to occupy grid positions that contained food. By trying out various actions and sequences of actions they learned a predictive model of the ways different actions tended to lead to different outcomes. Crucially, they learned the cue-seeking actions that delivered improved states of knowledge that would make their "optimistic predictions" (of finding the food) come true.

159 Rather than directly explore each upper arm: For the "simulated rat" experiments, see Friston, K., et al., "Active Inference and Epistemic Value," *Cognitive Neuroscience*, 2015; DOI: 10:1080/17588928:2015:1020053. See also Parr, T., and Friston, K. J., "Uncertainty, Epistemics and Active Inference," *Journal of the Royal Society Interface* Nov. 2017, 14(136):20170376). For another relevant treatment, see Tschantz A., Seth, A. K., and Buckley, C. L., "Learning Action-Oriented Models Through Active Inference," *PLoS Computational Biology* 16(4) (April 23, 2020): e1007805. For further links to active perception, see Parr, T., et al., "Perceptual Awareness and Active Inference," *Neuroscience of Consciousness* 29(1) (2019); DOI: 10:1093/nc/niz012.

161 But they are actually better at spotting: The work on learner drivers is reported in Land, M. F., and Tatler, B. W., *Looking and Acting: Vision and Eye Movements in Natural Behaviour* (Oxford: Oxford University Press, 2009).

162 This feeling of seamless integration: The exchange is reported in Gleick, J., *Genuis: The Life and Times of Richard Feynman* (New York: Vintage, 1993).

164 An increasingly familiar example can be found inside the human gut: See Furness, J. B., *The Enteric Nervous System* (Malden, MA: Blackwell, 2006). For a recent treatment, see Hibberd, T. J., et al., "A Novel Mode of Sympathetic Reflex Activation Mediated by the Enteric Nervous System," *eNeuro*. 2020 Aug 10;7(4):ENEURO.0187-20.2020.

164 For example, gut bacteria manufacture: A good introduction to this area is

Carpenter, S., "That Gut Feeling," in the *American Psychological Association's Monitor on Psychology* 43(8) (2012).

165 This showed that what looked like genetically determined: Both the experiments are reported in Bercik, P., et al., "The Intestinal Microbiota Affect Central Levels of Brain-Derived Neurotropic Factor and Behavior in Mice," *Gastroenterology* 141(2) (2011): 599–609.e3.

165 In the monkeys, stress-induced changes: See Bailey, M. T., et al., "Exposure to a Social Stressor Alters the Structure of the Intestinal Microbiota: Implications for Stressor-Induced Immunomodulation," *Brain, Behavior, and Immunity* 25 (2011): 397–407. See also Maltz, R. M., et al., "Social Stress Affects Colonic Inflammation, the Gut Microbiome, and Short-Chain Fatty Acid Levels and Receptors," *Journal of Pediatric Gastroenterology and Nutrition* 68(4) (2019): 533–540.

165 Life, as the philosopher of science John Dupré: See Dupré, J., and Malley, M. A. O., "Varieties of Living Things: Life at the Intersection of Lineage and Metabolism," *Philosophy Theory and Practice in Biology* 1:e003 (May 2009): 1–25.

167 This constant drip feed of directional information: See Josie Thaddeus-Johns's report "Meet the First Humans to Sense Where North Is," *The Guardian* (January 6, 2017).

167 In my previous work: See especially my 2003 book, *Natural-Born Cyborgs: Minds, Technologies, and the Future of Human Intelligence* (New York: Oxford University Press).

168 The philosopher Jerry Fodor once wrote: "If the mind happens in space at all": Fodor, J., "Diary," *London Review of Books* 21(19) (1999): 69.

168 Dave is now famous: See Chalmers, D., *The Conscious Mind* (Oxford University Press, 1996); and Chalmers, D., *Reality +: Virtual Worlds and the Problems of Philosophy* (Allen Lane, 2022).

169 But our short paper has become: The paper eventually appeared as Clark, A., and Chalmers, D., "The Extended Mind," *Analysis* 58(1) (1998): 7–19. Remarkably, Dave had commented to me, before it was published, that he thought our little piece had the potential to become a "modern classic." I considered this extremely unlikely. The paper was, I felt, just a kind of fun footnote to an emerging literature on embodied and distributed cognition.

171 Otto Goes to MoMA: The description of the extended mind argument that follows also draws upon a more recent version of that argument presented (by Clark and Chalmers) in de Cruz, H. (ed.), *Philosophy Illustrated* (Oxford University Press, 2020).

172 Our current view is that the true core: For this argument, see Chalmers, D., "Extended Cognition and Extended Consciousness," in Colombo, M., Irvine, E., and Stapleton, M. (eds.), *Andy Clark and His Critics* (Oxford University Press, 2019).

172 It is intriguing to note that a whole class: See Graves, A., et al., "Hybrid Computing Using a Neural Network with Dynamic External Memory," *Nature* 538, 471–476 (2016).

173 Despite having only a very tiny brain: See Zhang, S., et al., "Honeybee Memory: A Honeybee Knows What to Do and When," *Journal of Experimental Biology* 209(22) (November 15, 2006): 4420–4428.

173 Skin and skull do not: See Hurley, S., "The Varieties of Externalism," in Menary, R. (ed.), *The Extended Mind* (Cambridge: MIT Press, 2010).

174 If you insist that all that Otto: To respect the parity, some philosophers have even suggested that perhaps we should shrink the notion of what we currently believe to include only whatever we believe consciously in the here and now. But this is a weird way of respecting parity. Instead of embracing Otto as already having the belief that MoMA is on 53rd Street, you get to deny that but at the (too-high) cost of denying it to Inga too. See Gertler, B., "Overextending the Mind," in Chalmers, D. J. (ed.), *Philosophy of Mind: Classical and Contemporary Readings*, 2nd ed. (Oxford University Press, 2020).

174 But you can also impair my brain-based performances: I'm thinking here (to take just one example) of transcranial magnetic stimulation—a technique that can safely be used to modulate excitability levels in specific cortical areas, temporarily altering patterns of neuronal processing. See Valero-Cabré, A., et al., "Transcranial Magnetic Stimulation in Basic and Clinical Neuroscience: A Comprehensive Review of Fundamental Principles and Novel Insights," *Neuroscience and Biobehavioral Reviews* 83 (December 2017): 381–404.

174 As the range and use of assistive technologies: For a compelling description from a dementia sufferer, the literary editor Christine Lyall-Grant, see "My life Depends on Post-it Notes Now," by Victoria Lambert, *Daily Telegraph* (May 4, 2006). For a balanced take on the potential and the current limitations of assistive technologies, see Gibson, G., et al., "Personalisation, Customisation and Bricolage: How People with Dementia and Their Families Make Assistive Technology Work for Them," *Ageing and Society* 39(11), (2019): 2502–2519. And for something even more radical and challenging, try Robert Clowes, "The Internet Extended Person: Exoself or Doppelganger?" *Interdisciplinary Journal of Philosophy & Psychology* 15 (2020): 22.

175 Somehow, the canny biological brain: In *Supersizing the Mind*, I described the recruitment puzzle like this: "[the extended mind story] bequeaths a brand-new set of puzzles. It invokes an ill-understood process of 'recruitment' that soft-assembles a problem solving whole from a candidate pool that may include neural storage and processing routines, perceptual and motoric routines, external storage and operations, and a variety of . . . cycles involving self-produced material scaffolding [e.g., sketching]. And at its most radical, it depicts that process as proceeding without the benefit of a central controller." Clark, A., *Supersizing the Mind: Embodiment, Action, and Cognitive Extension* (New York: Oxford University Press, 2008), p. 137.

175 It is predictive brains, I believe: I speak here for myself, rather than for Chalmers. It's not so much that we disagree, but rather that Dave is less interested than I am in the whirrings and grindings of the neural machinery and keeps a closer focus on the behavioral-explanatory virtues. For his current thinking on the topic, again see Chalmers, C., "Extended Cognition and Extended Consciousness," in Colombo, Irvine, and Stapleton (eds.), *Andy Clark and His Critics*. The same volume contains a number of excellent critical treatments, along with a substantial reply.

176 Armed with those kinds of models or understanding: Among the key brain areas implicated in such complex processing, the human prefrontal cortex deserves a special mention. This large brain mass (which occupies around 10 percent of the volume of the human brain) has long been implicated in volition and higher cognitive function. Recently, a more unifying picture of its many cognitive roles has emerged. According to that picture, its primary function in cognitive control, learning, and memory is best understood as "anticipating prediction errors." This makes it a perfect tool for helping us to select and launch the right epistemic actions at the right time. See Alexander., W. H., and Brown, J. W., "Frontal Cortex Function as Derived from Hierarchical Predictive Coding," *Nature: Scientific Reports* 8(1) (2018): 3843.

180 Yet the alternative option (extending the mind): See Adams, F., and Aizawa, K., *The Bounds of Cognition*, 2nd ed. (Oxford: Blackwell, 2010); and Rupert, R., *Cognitive Systems and the Extended Mind* (Oxford: Oxford University Press, 2009). For more on the debate, see Colombo, Irvine, and Stapleton (eds.), *Andy Clark and His Critics*.

7. HACKING THE PREDICTION MACHINE

182 Similar results obtain for nausea: For pointers to key research papers in all these areas, a good place to start is with Price, D. D., Finniss, D. G., and Benedetti, F., "A Comprehensive Review of the Placebo Effect: Recent Advances and Current Thought," *Annual Review of Psychology* 59(1) (2008): 565–590. One of the first papers linking these effects to predictive processing accounts was Büchel, C., et al., "Placebo Analgesia: A Predictive Coding Perspective," *Neuron* 81(6) (2014): 1223–1239.

182 Looking outside of the medical context: For an excellent popular review, see Gary Greenberg's *New York Times Magazine* (November 7, 2018) piece, "What if the Placebo Effect Isn't a Trick?"

183 Those led to believe they were "enhanced": For a recent systematic review of the literature on sports performance and placebo, see Hurst P., et al., "The Placebo and Nocebo Effect on Sports Performance: A Systematic Review," *European Journal of Sport Science* 20(3) (April 2020): 279–292.

183 Such patients reported substantially more relief: The classic "sham surgery" result is Moseley J. B., et al., "A Controlled Trial of Arthroscopic Surgery for Osteoarthritis of the Knee," *New England Journal of Medicine* 347(2) (2002): 81–88. The study found that patients reported equal improvements in osteoarthritis pain regardless of whether they received a real surgical procedure or a sham. The results have been replicated and confirmed in various subsequent studies and meta-analyses, including Kirkley A., et al., "A Randomized Trial of Arthroscopic Surgery for Osteoarthritis of the Knee," *New England Journal of Medicine* 359(11) (2008): 1097–1107. See also Sihvone, R., Paavola, M., Malmivaara, A., and the FIDELITY (Finnish Degenerative Meniscal Lesion Study) Investigators, "Arthroscopic Partial Meniscectomy Versus Placebo Surgery for a Degenerative Meniscus Tear: A 2-Year Follow-up of the Randomised Controlled Trial," *Annals of the Rheumatic Diseases* 77 (2018): 188–195.

183 Honest (or "open-label") placebos: An excellent review of the "hon-

est placebo" research is Marchant, J., "Placebos: Honest Fakery," *Nature* 535 (2016): S14–S15. The work on irritable bowel syndrome appears as Kaptchuk, T. J., et al., "Placebos Without Deception: A Randomized Controlled Trial in Irritable Bowel Syndrome," *PLoS ONE* 5 (2010): e15591. For the work on cancer-related fatigue, see Zhou, E. S., et al., "Open-Label Placebo Reduces Fatigue in Cancer Survivors: A Randomized Trial," *Support Care Cancer* 27 (2019): 2179–2187. An accessible general account is Rich Hariday's "The 'Honest' Placebo: When Drugs Still Work Even Though Patients Know They're Fake," in the online magazine *New Atlas* (October 2018).

184 "Not only did we make it absolutely clear": The quoted comment is from an article by Rich Haridy published in the online technology, science, and news magazine *New Atlas* in October 2018—see https://newatlas.com/honest-placebo-treatment-research/56720/. The 2010 paper is Kaptchuk, T. J., et al., "Placebos Without Deception: A Randomized Controlled Trial in Irritable Bowel Syndrome," *PLoS ONE* 5(12): 2010: e15591.

184 But those in the latter (honest placebo) group: For this striking result, see Carvalho, C., et al., "Open-Label Placebo Treatment in Chronic Low Back Pain: A Randomized Controlled Trial," *Pain* 157(12) (2016): 2766–2772.

185 "even when administered openly": From Zhou, E. S., et al., "Open-Label Placebo Reduces Fatigue in Cancer Survivors: A Randomized Trial," *Support Care Cancer* 27 (2019): 2179–2187.

185 Contemporary thinking about placebo effects: This story forms a key part of the history of placebo research in the twentieth century. The doctor (Henry Beecher) went on to write a very influential short piece in 1955, called "The Powerful Placebo," appearing in *The Journal of the American Medical Association* 159 (1955): 1602–1606. A useful critical review of that piece is Kienle, G. S., and Kiene, H., "The Powerful Placebo Effect: Fact or Fiction?," *Journal of Clinical Epidemiology* 50(12) (1997): 1311–1318. For a comprehensive contemporary review, see Benedetti, F., *Placebo Effects*, 2nd ed. (Oxford: Oxford University Press, 2014).

185 Repeated administration of the actual (clinically effective) drug: See Headrick, J. P., et al., "Opioid Receptors and Cardioprotection—'Opioidergic Conditioning' of the Heart," *British Journal of Pharmacology* 172(8) (April 2015): 2026–2050. See also Corder, G., et al., "Endogenous and Exogenous Opioids in Pain," *Annual Review of Neuroscience* 41(1) (2018): 453–473.

186 After just four normally spaced genuine: See Benedetti, F., et al., "Teaching Neurons to Respond to Placebos," *The Journal of Physiology* 594(19) (2016): 5647–5660.

186 By training athletes using a performance-enhancing product: This possibility is raised in Chapter 5 ("Faster, Stronger, Fitter") of David Robson's excellent book *The Expectation Effect: How Your Mindset Can Transform Your Life* (Edinburgh: Canongate, 2022). That book is full of useful tips on how to exploit the eponymous "expectation effect" so as to improve our performance in a wide variety of ways, from fitness to diet, coping with stress, improving willpower, and enhancing problem solving.

187 "This concept of using precision medicine": The quoted comments were made by the study leader, Deepak Voora, in a *New Scientist* piece

(August 18, 2017) by Viviane Callier entitled "Genetic Test Helps People Avoid Statins That May Cause Them Pain." The source paper for the work is Peyser, B., et al., "Effects of Delivering SLCO1B1 Pharmacogenetic Information in Randomized Trial and Observational Settings," *Circulation: Genomic and Precision Medicine* 11(9) (2018): e002228.

188 Of special interest, as this science develops: It is possible, for example, that genetically determined differences play a role in this. The enzyme COMT (Catechol-O-methyltransferase), acting with other enzymes, helps determine the extent to which dopamine is metabolized in key brain areas. Those with higher levels of COMT metabolize more dopamine, thereby reducing its availability. These individuals also exhibit weaker placebo response. Lower levels of COMT have the opposite effect, seeming to promote placebo responsivity. This falls into place since dopamine is a key player in the complex (precision-weighted) balancing act that selectively weights predictions and sensory evidence. See Hall, K. T., Loscalzo, J., Kaptchuk, T. J., "Systems Pharmacogenomics—Gene, Disease, Drug and Placebo Interactions: A Case Study in COMT," *Pharmacogenomics* 20(7) (May 2019): 529–551.

188 Such individuals are experts: See Dienes, Z., et al., "Phenomenological Control as Cold Control," *Psychology of Consciousness: Theory, Research, and Practice* 9(2) (2022): 101–116; and Lush, P., et al., "Trait Phenomenological Control Predicts Experience of Mirror Synaesthesia and the Rubber Hand Illusion," *Nature Communications* 11, 4853 (2020). See also Martin, J. R., and Pacherie, E., "Alterations of Agency in Hypnosis: A New Predictive Coding Model," *Psychological Review* 126(1) (2019): 133–152.

189 But opioid treatments combined with VR: Neuroimaging results showed altered activity in the VR condition in key neural areas such as the insula and the thalamus, as well as in somatosensory areas. See Hoffman, H. G., et al., "The Analgesic Effects of Opioids and Immersive Virtual Reality Distraction: Evidence from Subjective and Functional Brain Imaging Assessments," *Anesthesia & Analgesia* 105 (2007): 1776–1783. A substantial popular piece introducing her work, as well as that of other leading pain theorists, is "A World of Pain" by Yudhijit Bhattacharjee in the January 2020 issue of *National Geographic*, pp. 46–69.

189 Soothing virtual reality scenes: See Tanja-Dijkstra K., et al., "The Soothing Sea: A Virtual Coastal Walk Can Reduce Experienced and Recollected Pain," *Environment and Behavior* 50(6) (2018): 599–625.

189 VR treatment has also been used successfully: On wound care in burns patients, see Hoffman, H. G., et al., "Virtual Reality as an Adjunctive Pain Control During Burn Wound Care in Adolescent Patients," *Pain* 85 (2000): 305–309. For a useful review confirming the efficacy of VR in burn injury care, see Malloy, K. M., and Milling, L. S., "The Effectiveness of Virtual Reality Distraction for Pain Reduction: A Systematic Review," *Clinical Psychology Review* 30 (2010): 1011–1018. See also Kipping, B., et al., "Virtual Reality for Acute Pain Reduction in Adolescents Undergoing Burn Wound Care: A Prospective Randomized Controlled Trial," *Burns* 38 (2012): 650–657. For the phantom limb work, see Ambron, E., et al., "Immersive Low-Cost Virtual Reality Treatment for Phantom Limb Pain: Evidence from Two Cases," *Frontiers in Neurology* 9 (2018): 67.

189 These subjective reports were further borne out: See http://www.hitl.washington.edu/projects/vrpain/.

189 But here, the results are mixed: For careful comparisons between music therapy and VR therapy, see Honzel, E., et al., "Virtual Reality, Music and Pain: Developing the Premise for an Interdisciplinary Approach to Pain Management," *Pain* 160 (2019): 9: 1909–1919.

189 "The V.R. segment in health care alone": See Ouyang, H., "Can Virtual Reality Help Ease Chronic Pain?," *New York Times Magazine* (April 26, 2022).

190 Designated as a Breakthrough Device by the FDA: See https://www.fda.gov/news-events/press-announcements/fda-authorizes-marketing-virtual-reality-system-chronic-pain-reduction.

190 This has since been confirmed: See Garcia, L. M., Birckhead, B. J, Krishnamurthy, P., et al., "An 8-Week Self-Administered At-Home Behavioral Skills–Based Virtual Reality Program for Chronic Low Back Pain: Double-Blind, Randomized, Placebo-Controlled Trial Conducted During COVID-19." *Journal of Medical Internet Research* (2021) 23(2):e26292. See also Garcia, L. M., Birckhead, B. J., Krishnamurthy, P., et al., "Three-Month Follow-Up Results of a Double-Blind, Randomized, Placebo-Controlled Trial of 8-Week Self-Administered At-Home Behavioral Skills–Based Virtual Reality (VR) for Chronic Low Back Pain," *Journal of Pain* (2021).

191 This could happen if the demonstrable: Caution is necessary because many conditions—including cardiovascular compromise, cancer, and viral and bacterial infection—will evolve and spiral unless identified and treated early. Proper diagnosis thus needs to be the essential precondition of responsible placebo use. See Benedetti, F., "The Dangerous Side of Placebo Research: Is Hard Science Boosting Pseudoscience?," *Clinical Pharmacology & Therapeutics* 106(6) (2019): 1166–1168.

193 Completing the prior self-affirmation task: See Martens, A., et al., "Combating Stereotype Threat: The Effect of Self-Affirmation on Women's Intellectual Performance," *Journal of Experimental Social Psychology* 42 (2006): 236–243. For a wide-ranging review, see Cohen, G. L., and Sherman, D. K., "The Psychology of Change: Self-Affirmation and Social Psychological Intervention," *Annual Review of Psychology* 65 (2014): 333–371.

194 Similarly, Black students in the U.S.: For the results with Black students in the U.S., see Cohen, G. L., et al., "Reducing the Racial Achievement Gap: A Social-Psychological Intervention," *Science* 313(5791) (September 1, 2006): 1307–1310. For the results with socioeconomically deprived children in the U.K., see Hadden, I. R., et al., "Self-Affirmation Reduces the Socioeconomic Attainment Gap in Schools in England," *British Journal of Educational Psychology* 90(2) (May 2020): 517–536. See also Goyer, J. P., et al., "Self-Affirmation Facilitates Minority Middle Schoolers' Progress Along College Trajectories," *Proceedings of the National Academy of Sciences of the United States of America* 114(29) (2017): 7594–7599.

194 Carefully chosen language can select: For a look at the many ways encounters with spoken words can impact predictive processing, see Lupyan, G., and Clark, A., "Words and the World: Predictive Coding and the Language-

Perception-Cognition Interface," *Current Directions in Psychological Science* 24(4) (2015): 279–284.

195 This work also showed that the effects: The breakfast omelet study was by Brown, S., et al., "We Are What We (Think We) Eat: The Effect of Expected Satiety on Subsequent Calorie Consumption," *Appetite* 152 (2010): 104717. The work on milkshakes and ghrelin is Crum, A. J., et al., "Mind over Milkshakes: Mindsets, Not Just Nutrients, Determine Ghrelin Response," *Health Psychology* 30(4) (2011): 424–429. For an accessible, evidence-led discussion of these (and many other) effects, see Robson, D., *The Expectation Effect* (Edinburgh: Canongate, 2022).

195 "an unpleasant sensory and emotional experience": The definition of pain is by the IASP (International Association for the Study of Pain), and is reported in Cohen, S. P., Vase, L., and Hooten, W. M., "Chronic Pain: An Update on Burden, Best Practices, and New Advances," *Lancet* 397(10289) (2021): 2082–2097. The definition of nociplastic pain that follows later is from the same source.

196 Nociplastic pain is thought to arise: See Fitzcharles, M. A., et al., "Nociplastic Pain: Towards an Understanding of Prevalent Pain Conditions," *Lancet* (2021 May) 29; 397(10289): 2098–2110.

196 Instead, there is a continuum of cases: See Freynhagen, R., et al., "Current Understanding of the Mixed Pain Concept: A Brief Narrative Review," *Current Medical Research and Opinion* 35 (2019): 1011–1018.

197 A similar profile applies: For some good discussion of the special case of chronic fatigue, see Nijs, J., et al., "In the Mind or in the Brain? Scientific Evidence for Central Sensitisation in Chronic Fatigue Syndrome," *European Journal of Clinical Investigation* 42 (2012): 203–212.

200 This can then have benefits outside the VR setting: See Trujillo, M. S., et al., "Embodiment in Virtual Reality for the Treatment of Chronic Low Back Pain: A Case Series," *Journal of Pain Research* 13 (2020): 3131–3137.

201 A few years ago, I came across an unusual: I was introduced to the case of Max Hawkins by Kathryn Nave, who was working with me on a large grant project on the predictive brain. We wrote about the case in a joint paper (along with other members of the grant team) and the next few paragraphs draw on that work. See Miller, M., et al., "The Value of Uncertainty," *Aeon Magazine*. https://aeon.co/essays/use-uncertainty-to-leverage-the-power-of-your-predictive-brain.

203 Deliberately engineering restricted forms of surprise: See Schwartenbeck, P., et al., "Exploration, Novelty, Surprise, and Free Energy Minimization," *Frontiers in Psychology* 4 (2013): 710. See also Domenech, P., Rheims, S., and Koechlin, E., "Neural Mechanisms Resolving Exploitation-Exploration Dilemmas in the Medial Prefrontal Cortex," *Science* 369(6507) (2020): eabb0184.

203 So much so, he said: The remark occurs in a *YPO Edge* 2019 talk entitled "Leaning in to Entropy," https://www.youtube.com/watch?v=3ecDsJrkKn4.

204 The last decade has seen a growing body: A sampling of specific papers includes: (on addiction): Bogenschutz, M. P., et al., "Psilocybin-Assisted

Treatment for Alcohol Dependence: A Proof-of-Concept Study," *Journal of Psychopharmacology* (Oxford, England), 29(3) (2015): 289–299; (on end-of-life distress): Ross, S., et al., "Rapid and Sustained Symptom Reduction Following Psilocybin Treatment for Anxiety and Depression in Patients with Life-Threatening Cancer: A Randomized Controlled Trial," *Journal of Psychopharmacology* 30(12) (2016): 1180; (on depression): Carhart-Harris, R. L., et al., "Psilocybin for Treatment-Resistant Depression: fMRI-Measured Brain Mechanisms," *Scientific Reports* 7(1) (2017): 13187. See also Barrett, F. S., Preller, K. H., and Kaelen, M., "Psychedelics and Music: Neuroscience and Therapeutic Implications," *International Review of Psychiatry* 30(8) (2018): 1–13.

204 Moreover, new research suggests that positive outcomes: See (for example) Kettner, H., et al., "From Egoism to Ecoism: Psychedelics Increase Nature Relatedness in a State-Mediated and Context-Dependent Manner," *International Journal of Environmental Research and Public Health* 16 (2019): 5147.

205 Brennan Spiegel, a leading proponent: The remarks are reported by Helen Ouyang, in "Can Virtual Reality Help Ease Chronic Pain?," *New York Times Magazine* (April 26, 2022).

206 Psychedelic drugs exert their strongest effects: Classic psychedelics all act as what are known as 5-HT2AR agonists—meaning that they bind to 5-HT2AR receptor sites and bring about their responses by that action. This is clear since blocking those receptors (e.g., with the antihypertensive drug ketanserin) extinguishes the psychedelic action of those molecules. The locations and densities of 5-HT2AR receptors in the neural architecture imply that the core effects of the drugs involve changes that occur in higher levels of cortical processing. See Carhart-Harris, R., "How Do Psychedelics Work?," *Current Opinion in Psychiatry* 32 (2019): 16–21.

207 Seminal neuroimaging (fMRI) work by Carhart-Harris: These decreases were not uniformly distributed, however. Instead, they occurred in key neural areas (thalamus, posterior cingulate cortex, medial prefrontal cortex) that seem to act as "hubs" orchestrating and coordinating activity across the whole brain. See Carhart-Harris, R. L., et al., "Neural Correlates of the Psychedelic State as Determined by fMRI Studies with Psilocybin," *Proceedings of the National Academy of Sciences* 109 (2012): 2138–2143.

207 But increased doses impact functioning: On ego dissolution under psychedelics, see Letheby, C., and Gerrans, P., "Self Unbound: Ego Dissolution in Psychedelic Experience," *Neuroscience of Consciousness* 2017(1) (2017): nix016; and for the specific results concerning posterior cingulate cortex, see Carhart-Harris, R. L., and Friston, K. J., "REBUS and the Anarchic Brain: Toward a Unified Model of the Brain Action of Psychedelics," *Pharmacological Reviews* 71 (2019): 316–344.

208 Properly informed, we can engineer psychedelic experiences: These issues are usefully discussed in section G of the REBUS paper (see previous note) by Carhart-Harris and Friston.

208 Such benefits would be especially marked: Not all recreational drugs share the potentially helpful profile of the classic psychedelics, and some (like MDMA and ketamine) share only part of it. Alcohol and many other "drugs

of addiction" act in a rather different way, seeming to hijack the brain's estimation of how well it is doing at reducing prediction error. Hijacking this process means that the brain gets fooled into estimating that it is doing far better than expected at reducing prediction error. Illusory success at minimizing large amounts of prediction error can then encourage a spiraling habit of use. This—working alongside multiple environmental factors—can make such habits especially hard to break. See Miller, M., Kiverstein, J., and Rietveld, E., "Brain and Cognition Embodying Addiction: A Predictive Processing Account," *Brain and Cognition* 138 (2020): 105495.

208 This is also one of the key effects of meditation: For some careful comparisons between the effects of different forms of meditation and of psychedelic drugs, see Millière, R., et al., "Psychedelics, Meditation, and Self-Consciousness," *Frontiers in Psychology* 9 (2018): 1475.

209 Focused-attention meditation: Intriguingly, neuroimaging work on meditation has found varying effects for various styles and forms of practice. For example, a 2016 meta-analysis of seventy-eight such studies found unique patterns of excitation and inhibition for several different techniques. But while each technique displayed its own unique fine-grained signature, they nearly all involved effects on the insular cortex—a neural region that—as we saw in Chapter 4—acts as a site at which multiple bodily (interoceptive) signals become integrated. See Fox, K. C. R., et al., "Functional Neuroanatomy of Meditation: A Review and Meta-analysis of 78 Functional Neuroimaging Investigations," *Neuroscience & Biobehavioral Reviews* 65 (2016): 208–228.

209 This means that even internal "information foraging": See Laukkonen, R. E., and Slagter, H. A., "From Many to (N)one: Meditation and the Plasticity of the Predictive Mind," *Neuroscience and Biobehavioral Reviews* 128 (April 2021): 199–217.

210 It is to gain better control over the precision-weighting performances: Despite some surface similarities, this kind of "stepping back from the self" is very different from the sometimes terrifying experiences reported by sufferers from depersonalization disorder (DPD)—a psychiatric condition sometimes brought about by abuse, torture, or extreme stress. In depersonalization disorder sufferers report a feeling of detachment and alienation from the world and from themselves—a feeling that has been described as deeply disturbing, amounting to a strong sense of personal nonexistence despite the presence of all your own memories and knowledge. Very occasionally, lifelong meditators can fall into such disturbed and disordered states. Indeed, depersonalization disorder of this kind has been famously described as "enlightenment's evil twin." But DPD is quite unlike normal meditation in that it involves a marked loss of personal control. For discussion of the commonalities and differences, see Deane, G., Miller, M., and Wilkinson, S., "Losing Ourselves: Active Inference, Depersonalization, and Meditation," *Frontiers in Psychology* 11 (2010): 539726. See also Gerrans, P., "Depersonalization Disorder, Affective Processing and Predictive Coding," *Review of Philosophy and Psychology* 10 (2019): 401–418.

210 By training attention and bodily awareness: There is a large and (it should be stressed) rather mixed literature on these kinds of effects. Good places to sample that literature include Farb, N., et al., "Interoception, Contemplative Practice, and Health," *Frontiers in Psychology* 6 (2015): 763; and Farb, N. A. S., Segal, Z. V., and Anderson, A. K., "Mindfulness Meditation Training Alters Cortical Representations of Interoceptive Attention," *Social Cognitive and Affective Neuroscience* 8 (2013): 15–26.

CONCLUSIONS: ECOLOGIES OF PREDICTION, POROUS TO THE WORLD

214 the so-called hard problem of explaining the nature: See Chalmers, D., *The Conscious Mind: In Search of a Fundamental Theory* (New York: Oxford University Press, 1996).

214 A set of misleading intuitions: See Dennett, D., *Consciousness Explained* (Boston: Little, Brown, 1991); and Frankish, K., "Illusionism as a Theory of Consciousness," *Journal of Consciousness Studies* 23(11–12) (2016): 11–39. For my own take on this, see my "Consciousness as Generative Entanglement," *Journal of Philosophy* 116(12) (2019): 645–662. See also Clark, A., Friston, K., and Wilkinson, S., "Bayesing Qualia: Consciousness as Inference, Not Raw Datum," *Journal of Consciousness Studies* 26(9–10) (2019): 19–33.

214 We need a much better understanding: As I write these words we have recently launched a new project devoted to just this topic. The project, funded by a European Research Council Synergy grant (XSCAPE—A New Methodology for the Study of Material Minds) brings together vision scientists, archaeologists, philosophers, and computational theorists, all working together to address the question of how different material and social environments impact our predictive minds.

215 The origins of our species' distinctive abilities: For some interesting speculations, see Deacon, T., *The Symbolic Species* (New York: Norton, 1997); Donald, M., *Origins of the Modern Mind: Three Stages in the Evolution of Culture and Cognition* (Cambridge: Harvard University Press, 1991); and Mithen, S., *The Prehistory of the Mind: A Search for the Origins of Art, Religion, and Science* (London: Thames & Hudson, 1996). See also Deacon, T. W., "Beyond the Symbolic Species," in Schilhab, T., Stjernfelt, F., and Deacon, T. (eds.), *The Symbolic Species Evolved*, vol. 6 of *Biosemiotics* (Dordrecht: Springer, 2012). See also Hutchins, E., "The Role of Cultural Practices in the Emergence of Modern Human Intelligence," *Philosophical Transactions of the Royal Society, B* 363(1499) (June 12, 2008): 2011–2019; and Hutchins, E., "The Cultural Ecosystem of Human Cognition," *Philosophical Psychology* 27 (2011): 34–49.

215 But however they arose, these skills: Symbolic culture on a human scale probably emerged thanks only to some lucky mosaic of minor adaptations, a set of historical contingencies, and the repeated "neural reuse" of resources that originally evolved to serve other purposes. For some compelling speculations, see Dehaene, S., *The Number Sense* (Oxford University Press, 1997); Smith, K., and Kirby, S., "Cultural Evolution: Implications for Understanding the Human Language Faculty and Its Evolution," *Philo-*

sophical Transactions of the Royal Society B, 363 (1509) (November 12, 2008): 3591–3603; and Heyes, C., *Cognitive Gadgets: The Cultural Evolution of Thinking* (Cambridge: Harvard University Press, 2018).

APPENDIX: SOME NUTS AND BOLTS

217 To keep the narrative flowing: Examples of such fuller treatments include Hohwy, J., *The Predictive Mind* (New York: Oxford University Press, 2013); and my own *Surfing Uncertainty: Prediction, Action, and the Embodied Mind* (New York: Oxford University Press, 2016). The view from cognitive neuroscience is elegantly captured by Anil Seth in *Being You: A Science of Consciousness* (Penguin, UK, 2021).

218 Similarly, a child who knows how: Notice how the child's use of the internal generative model for Lego design might be aided and abetted by exploratory action in the world. The child may push pieces around in space in ways that help them come up with new ideas about what structures are buildable. This kind of looping arrangement turns out to be very powerful, as we saw in Chapter 6.

218 To appreciate the power of a good generative model: The photorealistic images of "fake celebrities" were generated by the neural network described in Karras, T., Laine, S., and Aila, T., "A Style-Based Generator Architecture for Generative Adversarial Networks," *arXiv* 1812:04948 (December 2018): 1–12.

218 The multilevel artificial neural network: See Goodfellow, I., et al., "Generative Adversarial Nets," *Advances in Neural Information Processing Systems* 27 (2014): 2672–2680.

220 Instead, they must use observation-action sequences: For some early steps in this direction, see Çatal, O., et al., "Learning Generative State Space Models for Active Inference," *Frontiers in Computational Neuroscience* 14 (2020): 574372.

220 Moreover, this learning must be capable: See Parisi, G. I., et al., "Continual Lifelong Learning with Neural Networks: A Review," *Neural Networks* 113 (2019): 54–71.

223 All this happens extremely fast: Some of the most rapid early processing may (as noted in Chapters 1 and 5) also involve what's known—somewhat dauntingly—as "amortized inference." This provides a very direct, potentially ultra-rapid mapping from some forms of incoming sensory data to beliefs about the world. This could help set the scene for subsequent processes of iterative prediction and prediction-error-based refinement. See Tschantz, A., et al., "Hybrid Predictive Coding: Inferring, Fast and Slow," *arXiv* 2204:02169v2 (2022). See also Teufel, C., and Fletcher, P. C., "Forms of Prediction in the Nervous System," *Nature Reviews Neuroscience* 21(4) (2020): 231–242.

223 This means we perceive the "woods before the trees": For this broad picture, see Barrett, L. F., and Bar, M., "See It with Feeling: Affective Predictions in the Human Brain," *Philosophical Transactions of the Royal Society B*, 364(1521) (May 4, 2009): 1325–1334.

224 In this way action involves: See Friston, K. J., et al., "Action and Behavior: A Free-Energy Formulation," *Biological Cybernetics* 102 (2010): 227–260.

For an advanced application using a robotic platform, see Pio-Lopez, L., et al., "Active INFERENCE and Robot Control: A Case Study," *Journal of the Royal Society Interface* 13 (2016): 20160616.

225 This allows the brain to: See Friston, K., "The Free-Energy Principle: A Rough Guide to the Brain?, *Trends in Cognitive Sciences* 13 (2009): 293–301. See also Friston, K.,"Predictive Coding, Precision and Synchrony," *Cognitive Neuroscience* 3(3–4) (2012): 238–239.

225 Attention tends in this way: See Kok P., et al., "Attention Reverses the Effect of Prediction in Silencing Sensory Signals," *Cerebral Cortex* 22(9) (2012): 2197–2206.

226 This enabled my brain to resolve uncertainty: See Mirza, M. B., et al., "Introducing a Bayesian Model of Selective Attention Based on Active Inference," *Scientific Reports* 9(1) (September 2019): 13915.

227 This means that specific signals: Statistically speaking, "precision" names the reciprocal (the inverse) of the variance, where the variance is the estimated noisiness of some signal. See Feldman, H., and Friston, K., "Attention, Uncertainty, and Free-Energy," *Frontiers in Human Neuroscience* 2(4) (2010): 215.

227 As we go about our daily lives: For a closer look at the issue of levels and hierarchy in the brain, see Chapter 5 of my 2016 treatment, *Surfing Uncertainty: Prediction, Action, and the Embodied Mind* (New York: Oxford University Press).

227 In this way, variable precision-weighting: At the level of neurophysiology, there are many brain mechanisms working together to achieve this—especially complex neurotransmitter economies centered upon dopamine, serotonin, and acetylcholine. See, e.g., Friston, K., et al., "Dopamine, Affordance and Active Inference," *PLoS Computational Biology* 8(1) (2012): e1002327. See also Kanai, R., et al., "Cerebral Hierarchies: Predictive Processing, Precision and the Pulvinar," *Philosophical Transactions of the Royal Society B*, 370(1668) (May 19, 2015): 20140169. For a fairly comprehensive review of the empirical evidence for predictive processing more generally, see Walsh, K. S., et al., "Evaluating the Neurophysiological Evidence for Predictive Processing as a Model of Perception," *Annals of the New York Academy of Sciences* 1464(1) (2020): 242–268.

228 The attempt to minimize those errors: For a thorough introduction to this whole picture, from the conceptual landscape all the way to the neurophysiology and with a handy toolkit for simulations, see *Active Inference: The Free Energy Principle in Mind, Brain, and Behavior,* by Thomas Parr, Giovanni Pezzulo, and Karl J. Friston (Cambridge: MIT Press, 2022).

Index

(Page references in *italics* refer to illustrations.)